BUILDING A
SENSORY
PROGRAM

A BREWER'S GUIDE TO BEER EVALUATION

"Whether you are firing up your brewhouse for the first batch or producing well over 100,000 barrels of beer a year, Pat Fahey's *Building A Sensory Program* belongs in your collection of brewing books. Building a culture of sensory is crucial to every brewery's success, and this book will help you succeed. Speaking as someone who has started a sensory program at a brewery, this book will not only teach you the basics but will carry you years into the development of your sensory program."

—RILEY SEITZ,
QA/QC Manager, Surly Brewing Company

"Building a quality sensory program at a brewery allows for a significant protection of the product. It aids in building the best profiles while allowing a brewery to ensure consistency throughout brewing. *Building a Sensory Program* is an excellent resource to take those first steps in creating quality products and maintaining them."

—ANNETTE FRITSCH,
Senior Director of Product Development and Sensory, Boston Beer Company

"The unerring ability to deliver to your customers exactly the beer that you promised them is the hallmark of the truly professional brewer. Only a solid and reliable sensory program can guarantee a brewery's ongoing success at brewing to target and making sure its beers are as delicious as the brewer hopes they are. Medals are cool, Instagram cred is great, but here Pat Fahey shows you how to ensure and track The Right Stuff—actual quality."

—GARRETT OLIVER,
Brewmaster, The Brooklyn Brewery
Editor-in-Chief, *The Oxford Companion to Beer*
Michael James Jackson Foundation for Brewing and Distilling

"While sensory science has always been linked with the brewing industry, its practicability within small- to medium-sized operations has not been adequately addressed until now. Pat Fahey has filled a gaping hole in sensory literature by providing information and guidance to people and programs that do not have the luxury of extra time, resources, institutionalized knowledge, or advanced statistical backgrounds. The first step to building a sensory program is to taste the beer, the second step should be to read this book."

—CRAIG THOMAS, MSc,
Sensory Research Analyst, Firestone Walker Brewing Company

"The industry has desperately needed *Building a Sensory Program* for quite some time now and we should be excited that the details and approach of modern, practical, and scientifically grounded beer sensory can be accessed in a single place. Folks new to beer sensory will save months of self-directed research into dozens of unrelated sources and will understand the purposes and capabilities of a sensory program. Current panel administrators now have a volume of immediately useful information at their fingertips, and will find more than few new tips and methods to try out!"

—BENNETT THOMPSON,
Quality Manager, Half Acre Beer Company

"When I was starting the sensory program at New Belgium in 1999, there wasn't a craft beer sensory community or applicable technical resources out there. So, while I was full of determination and enthusiasm, it was sometimes a lonely and uncertain endeavor. Pat's book, *Building a Sensory Program: A Brewer's Guide to Beer Evaluation*, is the culmination of the incredible progress that's been made since then and will be every sensory leader's best friend and guiding light."

—LAUREN LIMBACH,
Wood Cellar Director & Blender, New Belgium Brewing

"Most successful breweries I know have a very strong sensory program and consider it the most important tool in their quality program. Pat Fahey's book covers everything you need to know about sensory analysis and how to build a sensory program. He relies on research and industry experts to dissect the biochemistry of the senses of taste and smell, and then lays out an extensive, detailed plan for building a top-quality sensory program that breweries of all sizes and types can use to improve their beer quality and consistency. It is a great read and a wonderful resource."

—MITCH STEELE,
COO and Brewmaster, New Realm Brewing Company

"A sensory program is essential to every brewery, no matter its size. In this book, Pat Fahey brings together a wealth of sensory knowledge that will positively impact the beer industry and beer quality. This book is a wonderful reference!"

—AMAEY MUNDKUR,
Program Director, American Society of Brewing Chemists

"Starting or expanding a sensory panel can be costly and challenging for a brewery without clear guidance. Pat Fahey's *Building a Sensory Program* is a comprehensive resource that clearly explains sensory science and the details necessary for assembling a successful panel."

—KARL ARNBERG,
Sensory Program Manager, Allagash Brewing Company

BUILDING A
SENSORY PROGRAM

A BREWER'S GUIDE TO BEER EVALUATION

BY PAT FAHEY

BREWERS PUBLICATIONS®

Brewers Publications®
A Division of the Brewers Association
PO Box 1679, Boulder, Colorado 80306-1679
BrewersAssociation.org
BrewersPublications.com

Proudly Printed in the United States of America.
10 9 8 7 6 5 4 3 2 1
ISBN-13: 978-1-938469-67-1
ISBN-10: 1-938469-67-4
EISBN: 978-1-938469-68-8

Library of Congress Cataloging-in-Publication Data

Names: Fahey, Pat, 1987- author.
Title: Building a sensory program : a brewer's guide to beer evaluation / by Pat Fahey.
Description: Boulder, Colorado : Brewers Publications, [2021] | Includes bibliographical references and index. | Summary: "Human flavor perception is incredibly complex and impacts daily decision making in the brewery. Building a Sensory Program discusses sensory systems, sources of bias, tasting techniques, required equipment, taster training, and panel maintenance. Learn about different testing methods, data use, and how to use sensory to adjust recipes and build or blend new beers. This book provides the reader a bedrock for an intelligently designed quality sensory program"-- Provided by publisher.
Identifiers: LCCN 2020054988 (print) | LCCN 2020054989 (ebook) | ISBN 9781938469671 (trade paperback) | ISBN 1938469674 (trade paperback) | ISBN 9781938469688 (ebook)
Subjects: LCSH: Beer tasting. | Sensory evaluation. | Program management.
Classification: LCC TP577 .F34 2021 (print) | LCC TP577 (ebook) | DDC 663/.42--dc23
LC record available at https://lccn.loc.gov/2020054988
LC ebook record available at https://lccn.loc.gov/2020054989

Publisher: Kristi Switzer
Technical Editor: Lindsay Barr, Karl Arnberg
Copyediting: Iain Cox
Indexing: Doug Easton
Art Director, cover and interior design, and production: Jason Smith
Production: Justin Petersen

To Averie.

TABLE OF CONTENTS

FOREWORD

Pat Fahey and I first crossed paths in the winter of 2013 when he popped into my office at New Belgium to introduce himself as Content Director of the Cicerone® Certification Program. Cicerone had just started making waves in the brewing industry, and I was interested to learn how Pat was framing the task of how to unravel the obscurities of beer flavor. During our conversation I discovered that Pat is an active observer, capable of honing his creativity to find unconventional solutions to big questions. These characteristics made him uniquely qualified to tackle his work with Cicerone and, later, to write this book.

We talked for hours in my office, working through concepts that had only just started bubbling to the surface in our minds. This initial meeting set the stage for what would become a meaningful friendship that, to this day, encourages challenging norms, identifying gaps in our knowledge, and pulling at the threads that our questions revealed. Throughout our friendship, we recognized our many overlapping philosophies: systems are empowering, sensory should be accessible, and value must lead. These philosophies became the fabric with which this book was crafted.

SYSTEMS ARE EMPOWERING

Pat is an active observer. Anyone who has dined with him knows this curiosity first-hand when he hands you the most esoteric item from the menu for the sheer purpose of sharing a unique experience. He is not a solo explorer. He invites you to join in his curiosity through his practiced ability to communicate experiences, and this is what makes him such a gifted educator. In writing this book, Pat distills complicated concepts into digestible nuggets, synthesizing the particularities of existing sensory methods and transforming them into useful systems that can be applied regardless of your available resources.

Systems empower small businesses by fast-tracking the process of arriving at answers to frequently posed questions. In sensory, we need to know what to make, if a batch is suitable for release, and how to tweak the process to accomplish the desired outcome. By applying systems to these questions consistently, our data works for us and the process of arriving at an answer is fast and efficient. Imagine the burden of having to consciously decide how to put on your pants every morning. If we constantly changed the process of achieving "pants on," it would take hours to get ready each morning, potentially with questionable results. In this book Pat highlights the methods at the core of every successful sensory program and outlines their processes of implementation. Once this foundation is built, you will find you have the time and energy to creatively expand your program where it makes the most impact for your business.

SENSORY SHOULD BE ACCESSIBLE

Pat is a creative. He unknowingly put this trait on display when one winter visit to Fort Collins he surprised me by stopping mid-stride on our way to dinner to seat himself at a public piano in the middle of town. I started to sigh and roll my eyes, thinking I was in for a goofy rendition of "Mary Had a Little Lamb," but was amazed when he nonchalantly began playing Rachmaninoff's Prelude in C-sharp Minor. Musicians can see patterns, improvise, and elegantly connect a story from start to finish. In brewing, Pat has an intuitive understanding of the process and a keen awareness of the quality system's interconnectedness. In sensory, he is capable of clearly defining questions and utilizing available resources and systems to find solutions. Pat and I both believe sensory should be accessible, and we have made it our mission to democratize the techniques that have long been employed by those with abundant resources.

The senses are complex, and humans are messy instruments. The sensory industry's answer to this human variability has been to achieve high panel numbers and hammer in countless hours of training. These requirements have made sensory intimidating and inaccessible to small businesses. In this text, Pat takes a pragmatic approach to sensory by focusing on achieving meaningful results while properly framing the limitations of each data source. He spends time outlining sources of bias and detailing common pitfalls so you can build best practices into your program and achieve actionable tasting results.

VALUE MUST LEAD

Pat takes an unconventional approach to sensory and education. If you are looking for a formulaic academic text, this is not the book for you. Rather than painstakingly describing each sensory method, he gives practical guidance for starting a sensory program. Pat is able to find the balance between giving thorough explanations while leaving plenty of room for your creativity. By giving weight to the "why," this book helps simplify the sensory process so your energy can be directed to where it will have the most impact.

To date, no sensory text quite like this exists. While several books and articles have been published, this is the first to directly focus on the application of sensory science for small businesses, specifically craft brewers. You will learn how to frame questions, the methods to apply to your objectives, and the decisions you can make when employing flavor analysis.

Building a Sensory Program brings the value of a sensory program's clearly into view. It will be both instrumental in starting your own program and a useful reference to keep on hand for years to come.

Lindsay Barr
Founding Partner and CSO
DraughtLab Sensory Software

ACKNOWLEDGMENTS

I wanted to avoid opening with a trite statement about how it takes a village, but the truth is that writing a book is never a solitary venture. While I put all of the words down on the page, countless others helped shape my sensory knowledge, views, and philosophy. Through their generous help I was able to bring this project to fruition and I owe them all a debt of gratitude.

First, I want to thank those who helped me get to where I am today. My parents, who have always supported me no matter what path I chose, even if they do have to suffer through trying to explain my job to their friends. Ray Daniels, for taking a chance on a 24-year-old kid and officially bringing me into this tight-knit industry that I now call home. A wide array of colleagues, mentors, and friends who have both guided me and helped open doors along the way, including (but not limited to) Adam Schulte, Amaey Mundkur, Brian Dewald, Charlie Franklin, Chris Quinn, Chuck Skypeck, Corey Shovein, Dave Kahle, Donn Bichsel Jr., Drew Larson, Evan Price, Gera Exire LaTour, Jamie Lash, Jake Lewis, Jason Pratt, Jim Randall, John Mallett, Julia Herz, Karen McVicker-Black, Lanny Hoff, Neil Witte, Nicole Garneau, and Randy Mosher. Brett Steele, Chris Pisney, and Shana Solarte for reviewing my initial proposal and so much more, and the rest of my co-workers at Cicerone, for putting up with me and helping me stay sane throughout this project.

An enormous thank you to all of the sensory professionals who welcomed me into their breweries and/or took the time to discuss their programs with me at length. Everyone I interacted with far exceeded my expectations in their willingness to give their time and share their experience and accumulated knowledge. Thank you Alex Hazelmyer, Ali Schultz, and Jacob Fuentes (New Belgium Brewing Company); Amanda Milford (Cigar City Brewing); Amy Noga, Micaela Kraft, and Steve McCarthy (Boston Beer Company); Andrew Heyboer and Eric Adkins (Breckenridge Brewery); Bennett Thompson (Half Acre Beer Company); Craig Thomas (Firestone Walker Brewing Company); Karl Arnberg (Allagash Brewing Company); Kevin Payne and Lauren Torres (Bell's Brewery); Kristen Soucek (Revolution Brewing); Meghan Peltz (Sierra Nevada Brewing Company); Bob Galligan and Riley Seitz (Surly Brewing Company); Riley Wetzel (The Bruery); and Stephen McIngvale and Sue Thompson (MillerCoors).

Thanks to those who offered their expertise in addressing specific topics—in particular Chris White and Erik Fowler of White Labs for providing sensory techniques for assessing yeast and Richard Preiss of Escarpment Labs for sharing a wealth of information related to tetrahydropyridine (THP). Additionally, a special thank you to Bill Simpson of Cara Technology and Aroxa™ for helping to deepen and enrich my personal flavor knowledge and capabilities over the years, and for graciously helping me address any flavor-related questions that I had while researching, writing, and editing this book.

Thank you to the many brewers and blenders that contributed to the chapter on blending wood-aged beer. The beer industry truly cultivates an openness and fellowship not found anywhere else. I appreciate the insight and precise detail each of them provided into their unique programs; I only wish I could have written more about the fascinating techniques involved in producing these sorts of beers. The passion they showed for both their beers and their process was infectious, and our conversations energized and invigorated me as I pushed toward completing book. Thank you to Andrew Zinn (Wicked Weed Brewing), Andy Parker (Avery Brewing Company), Averie Swanson (Keeping Together), Blake Tyers (Creature Comforts Brewing), Cole Hackbarth (Rhinegeist Brewery), Cory King (Side Project Brewing), Frank Boon (Brouwerij Boon), Harrison McCabe (Beachwood Blendery), Jason Perkins (Allagash Brewing Company), Jay Goodwin (The Rare Barrel), Jeremy Grinkey (The Bruery), Jeremy Inzer (Fonta Flora Brewery), Jim Crooks (Firestone Walker Barrelworks), Lauren Limbach (New Belgium Brewing Company), Marty Scott (Revolution Brewing), Mike Fava (Oxbow Brewing Company), Ron Jeffries (Jolly Pumpkin Artisanal Ales), and Willem Van Herreweghen (Timmermans). And an additional heartfelt thank you to Lauren Limbach for contributing an image of her foeder tasting notes to the book.

As further testament to the comradery of this industry, thank you to those that helped make introductions or connections, including Adam D'Antonio, Chris Shields, Chris Swersey, Ken Smith, and Neil Callaghan.

Thank you to the entire Brewers Publications team and in particular to Kristi Switzer for your patience and encouragement throughout this project. At times it seemed to stretch on forever, but your reassuring presence helped guide me through to the end. Thank you to my copyeditor, Iain Cox, for improving the content and clarity of the book through your incisive questions, and for sharpening my writing all around. And thank you to Karl Arnberg—in addition to helping source images and providing an enjoyable and thorough description of your program at Allagash, your technical edit helped address some missing pieces in the book and made for a stronger finished product.

An extra special thank you to Lindsay Barr of DraughtLab, without whom this book would not exist. Yes, you served as my technical editor, but you also acted as my sensory guru for the entirety of the project, from helping me shape my outline, to countless visits, discussions, and phone calls to clarify details and discuss sensory philosophy, and on through multiple rounds of edits and revisions. Lindsay, I honestly cannot thank you enough for the time and energy you invested in helping me with this book—you are a rock star and I owe you at least one very expensive dinner.

And finally, thank you to my amazing partner, Averie Swanson. You offered help in the form of feedback, introductions to colleagues, and your own expertise in the realm of wood-aged beer. However, above all you supported me throughout this project, through overseas adventures and cross-country road trips spent writing, through late nights and long weekends spent huddled over my computer editing instead of spending time together, all with nary a complaint.. Thank you for sticking by my side throughout this. endeavor—now we celebrate!

INTRODUCTION

Beer quality has never been more important. In the US alone, the number of operating breweries has nearly quintupled in the last decade, from 1,813 in 2010 to 8,363 in 2019.[1] Consumers are inundated with options when it comes to choosing what beer to drink. As the number of breweries has grown, so too has consumer knowledge and expectations. In the face of increasing competition, quality no longer serves as a point of differentiation but has become an imperative.

Most breweries today understand that an effective sensory program would help them deliver quality beer to market. Yet many breweries do not have any sort of sensory program in place, either formal or informal. Resource allocation is a constant consideration in small and even mid-sized breweries, with each department vying for access to a limited pool of capital and time. Some breweries resign themselves to investing in a sensory program further down the road when they have more time and money. But putting off developing a sensory program virtually ensures that you will not get around to establishing a program until it is too late. If you are lucky, your wake-up call may come in the form of consumer complaints on a brand that has drifted away from its original target. If you are unlucky, you may find yourself responding to an entirely preventable, catastrophic product recall.

However, even breweries that want to begin a sensory program often suffer from a lack of direction. Through no fault of their own, most breweries have no idea where to start in building a program, and to date very few resources exist to assist in the task. Current sensory texts are designed for large, industrial manufacturers of consumer products, and describe best practices for tests involving a minimum of 20–30 trained tasters. This type of sensory program is out of the reach of all but a handful of very large breweries. Furthermore, these texts often place undue emphasis on difference testing (e.g., triangle tests), which can answer questions related to process adjustments or recipe changes but are not well suited for tackling quality control. Without guidance, some breweries attempt to apply these types of tests to product release or shelf life assessment, often with unsatisfactory results.

This book aims to fill that resource gap. My goal with this text is to combine the best practices found in sensory textbooks with the techniques used by sensory leaders in the beer industry to yield a set of good practices that small and medium-sized breweries can actually implement. A sensory program at a small brewery can yield useful, actionable data with as little time investment as a couple of hours each week. No brewery is too small—as long as you have more than one person tasting, you can hold viable tasting sessions. Your sensory program will serve as a tool to help you answer questions about your beer, such as whether your beer is consistent, how long your beer should remain in the market, and what new beer you should make next. By better understanding the questions that you

---

[1] "National Beer Sales & Production Data," Brewers Association (website), Stats and Data, accessed August 15, 2020, https://www.brewersassociation.org/statistics-and-data/national-beer-stats/.

want answered, you will design a thoughtful and intentional sensory program that delivers useful results from day one. Once you begin to reap the rewards of your program you may decide to invest more time and resources, but remember that your program will continue to pay dividends even if you only devote a small amount of time to it.

I have divided this book into two main portions: theory and practice. Chapters 1–8 cover information on sensory systems, bias, tasting techniques, the different types of tests available to your panel, and equipment needed. These chapters will help inform the way you set up your panel and train your panelists. You certainly do not need to master this information to be an effective panel leader, but these chapters should serve as a reference for any questions that you have along the way. Chapters 9–13 cover the practical elements of running an effective sensory program, beginning with panelist

training before moving into the wide variety of tasks you can accomplish with your panel, from product release and shelf life assessment to recipe design and brand building. Chapter 14 examines the art of blending wood-aged beer, focusing on how different brewers use sensory evaluation to guide their decision-making.

Your journey into the vast and thrilling world of sensory exploration begins today. While you may have been drawn here with the sole objective of establishing a sensory program, you will also gain insight into the way your own palate works, learn how to motivate and excite your panelists, and develop a more intimate understanding of your beers. By purchasing this book, you have already shown that you are dedicated to building sensory evaluation into the culture of your brewery. Thank you for taking this first step with me—I wish you success in the never-ending quest for better beer.

1

THE VALUE OF A SENSORY PROGRAM

To the uninitiated, the task of building a sensory program may appear daunting. Setting aside the logistical considerations for now, you might find the idea of applying formal sensory evaluation to beer somewhat intimidating; many people certainly do. Even if you do not have concerns about your own sensory abilities, you may wonder how on Earth you will be able to recruit an entire panel of capable tasters for your program. There seems to be a pervasive belief that skilled tasters possess an innate talent for tasting, which, while true for a minority of individuals, is typically not the case. Many people lack confidence in their personal sensory abilities. In training sessions, I have heard the refrain "I'm just not a very good taster" all too often. But do not fear, sensory skills are highly trainable, and making decisions based on your senses is a lot less intimidating than it may initially appear. In fact, you probably already make multiple sensory-based decisions every day.

Throughout human history, the sensory qualities of aroma and taste have driven perceptions of both quality and value. For thousands of years, sensory assessment has formed the cornerstone of consumer decisions relating to a wide variety of consumable products. In ancient Egypt, Rome, and Persia, fine-smelling perfumes, incense, and spices commanded lofty prices (Doty 2015, 4). Since the earliest days of trade, consumers would evaluate a merchant's products using their senses, purchasing based on the quality of aroma or taste. While consumer products have certainly evolved, our evaluation methods remain constant. You would probably be surprised by the multitude of minor decisions you make using sensory assessments. Do I like the way this soap smells? Do I enjoy the way this beverage tastes? Does this soup need more salt? These simple judgments form the basis of our preferences and purchasing habits. While manufacturers of consumer products pay close attention to consistently meeting a set of specifications, marketers and advertisers alike often focus on the sensory experience, given that it drives so much of our behavior.

Beer is no different. Brewers use specifications like original gravity and IBUs to help hit a target profile, but the ultimate consideration is the sensory experience. Does the beer taste good? Do consumers find it appealing? These are the questions that drive creation of new beer brands. These are the qualities that make for excellent beer.

Beer can be measured and analyzed in innumerable ways. In addition to simple readings of gravity or pH, complex instruments can be used to measure color, bitterness, and alcohol level. Gas chromatography–mass spectrometry (GCMS) can be used to separate out a complex mixture like beer into its individual chemical constituents, allowing a lab technician to measure the exact levels of compounds present (e.g., diacetyl). High-tech "electronic noses" have been developed in recent years, allowing researchers to pick out flaws or aging characteristics with a high degree of accuracy (Ghasemi-Varnamkhasti et al. 2011, 57).

However, none of these analytical instruments can suitably replicate the human tasting experience. While the following statement may seem obvious, it is worth stating explicitly: the only way to really know how your beer tastes is to taste it.

THE GOAL OF YOUR SENSORY PROGRAM

The goals of a brewery sensory program should support the goals of the brewery itself. I would argue that all breweries share a central, primary goal, which is to release beer that tastes the way they want it to. Thus, a new sensory program should concentrate first and foremost on product release. By starting with a focus on beer release, your sensory program will offer a powerful quality check on your outgoing beer and your beermaking process.

QUALITY CONTROL

Quality is not a new concept. Humans seek quality in virtually all products, from the food they eat to the clothes they wear. While each product category may be judged using different metrics, the notion of quality always plays a role. Historically, skilled artisans and craft guilds drove consumer perceptions of quality—a well-made pair of shoes might carry the mark of a renowned shoemaker, while a fine Burgundy might bear the appellation of a specific region known for a certain type of wine. The reputation of the artisan or region conveyed an impression of quality, guaranteeing that the product would be up to snuff.

Then came the Industrial Revolution, entirely transforming the concept of craftsmanship. Factories became the norm, with craftsmen working for a manager or supervisor who would audit the quality of the products. An obsession with efficiency led to the development of the "scientific management" process in the US. Pioneered by industrial engineer Frederick Winslow Taylor, scientific management broke complex jobs into simple, repetitive steps (Kolb and Hoover 2012, 7). Productivity went up and prices came down, but, with the diffusion of responsibility, quality suffered.

It was under these conditions that a scientific approach to quality control was born. The early twentieth century saw the development of statistical quality control methods. Some of these methods were developed at breweries, such as the Student's t-test, developed by a chemist at Guinness (see p. 84) (Dodge 2008, 234). Over the course of the century the value of quality standards in manufacturing became more apparent, leading larger manufacturing organizations to build out entire departments whose sole purpose was to ensure quality in production.

Indeed, megabreweries invest tremendous resources in quality control for their products. Small breweries, by contrast, might be lucky to even have a single employee dedicated to quality. More often than not, if quality control is considered at all, it is considered to be the responsibility of everyone at the organization, a solution that usually means no one is explicitly focused on quality. That does not have to be the case though! As we will explore in the coming chapters, there is much that a small brewery can do to improve the quality of its beer using resources already at its disposal. However, before we go any further talking about quality or quality control, we should make sure that we are starting from a shared understanding of what quality actually is.

Quality Defined

It may surprise you if you have never thought about it before, but it is actually somewhat difficult to pin down a working definition of quality. If you consult texts on quality control, you will find a slew of different attributes that can be used to define quality. Some focus on the need for a quality product to conform to a set of specifications, some aim to achieve quality through reducing overall product variability, while others address usability (Muñoz, Civille, and Carr 1992, 3). Others merely refer to the "fineness" or excellence of the product—vague descriptions that at best are highly subjective and at worst are quite useless.[1] The best definitions of quality typically revolve around how well a given product satisfies the needs of its consumers. In the realm of food and beverages, this type of definition seems particularly apt—a quality beer should please and delight those that drink it.

The Brewers Association Quality Subcommittee defines quality beer as "a beer that is responsibly produced using wholesome ingredients, consistent brewing techniques and good

---

manufacturing practices, which exhibits flavor characteristics that are consistently aligned with both the brewer's and beer drinker's expectations."[2] Given that the subcommittee's definition encapsulates multiple elements of the product cycle and focuses on meeting the needs of the consumer, it seems as good a foundation as any. With that in mind, we turn to quality control, which focuses specifically on monitoring and achieving consistency in your beer.

Instrumental Quality Control

In many industries, instrumental measurements and metrics form the core of a strong quality control program. The beer industry is no exception—breweries large and small rely on instrumental measurements to ensure consistency in their products. A robust brewery quality control program will utilize chemistry, microbiology, and sensory methods to track the quality of their beers.

Fortunately, the most important instruments for brewers also happen to be some of the most affordable and accessible. Brewers use hydrometers and pH meters to measure key metrics during wort production and fermentation. Similarly, brewers can manage pitching rates by using microscopes and hemocytometers to perform cell counts. These basic instrumental measurements can greatly improve a brewery's batch-to-batch consistency.

Instruments have significant advantages over human subjects in several areas. Instruments can perform hundreds of measurements to a high level of precision without becoming fatigued. Also, because the instrument simply reports data without interpretation, instruments do not suffer from biased judgment. When attempting to measure a single variable or metric, instrumental methods outperform human subjects every time.

However, instruments do come with notable downsides. Most analytical instruments are quite expensive, a prime consideration for all but the largest breweries. The instruments mentioned above—hydrometers, pH meters, microscopes, and hemocytometers—are well within the budget of all breweries and should be considered essential equipment for proper beer making. However, more advanced equipment, such as an alcolyzer or a GCMS, can cost tens or even hundreds of thousands of dollars. While this sort of equipment can certainly provide useful data, you can secure the same data by sending beer out for analysis if necessary. Also, some of this information is nice to know rather than need to know and, as such, can ultimately be forgone by the small brewer.

Beyond cost, there are several other areas in which instruments fall short of human subjects. While instruments offer unparalleled precision, in certain cases some human senses—most notably smell—have a higher level of sensitivity. For example, human subjects can perceive volatile thiols such as 4MMP (4-mercapto-4-methylpentan-2-one)—an important flavor marker in certain hop varieties—in single-digit parts per trillion, beyond the sensitivity threshold of most instruments (Rettberg, Biendl, and Garbe 2018, 6).

Even more critically, instruments only measure exactly what you tell them to. When collecting instrumental measurements, you must make sure you are collecting the correct data, as the instrument will only report the specific variable you are targeting when taking a measurement. Beer color measurement offers an instructive example of this. For beers proceeding normally along the color spectrum from straw to gold, amber, brown, and eventually black, Standard Reference Method (SRM) or European Brewing Convention (EBC) values allow us to accurately predict the color of the beer. However, if the beer exhibits a reddish color due to the use of crystal malt or roasted barley, the resulting color defies quantification using these methods. (This can also occur if using fruit or other adjuncts.) While the difference is obvious to the naked eye, an instrument will miss the distinction since it falls outside of what we told the instrument to look for (Smythe and Bamforth 2000, 166). Within human flavor perception, our senses analyze all stimuli present, leaving open the possibility of recognizing an unexpected flavor compound or other beer characteristic that may be missed through instrumental analysis.

Most importantly, however, is that no instrument currently exists that can replicate the way humans perceive flavor. The human flavor experience is incredibly complex. When we taste beer, our brains instantly combine sensory input from each of our different sensory modalities (i.e., taste, aroma, sight, touch,

---

2   Chuck Skypeck, "BA Quality Subcommittee Sets Vision, Mission, Definition of 'Quality Beer,'" Brewers Association (website), Association News, January 29, 2015, https://www.brewersassociation.org/association-news/ba-quality-subcommittee-sets-vision-mission-definition-quality-beer/.

and sound). This complex combination of inputs is then referenced against our personal experiences and memories before returning as a specific flavor perception. Instruments cannot replicate this process for a number of reasons. Firstly, instruments cannot reproduce the physical filtering that occurs in our saliva and mucus membranes to selectively determine which compounds actually make it to our receptors (Lawless and Heymann 2010, 30). Secondly, instruments cannot map the natural perceived intensities of aromas and tastes. Instruments measure absolute concentrations of taste and aroma molecules, but do not inherently know whether a given compound is perceptible to humans in the parts per thousand range or the parts per trillion range. Finally, instruments lack the ability to replicate the interpretation of sensory stimuli performed by the brain. When we taste a beer, the brain does not simply receive a collection of inputs and report back data as an instrument would. Instead, various portions of our sensory pathways interpret and process the stimuli en route to the neocortex. The neocortex then maps interactions across the different senses while filtering and ignoring certain stimuli to ultimately synthesize flavor (Shepherd 2012, 110). Given that we only partially understand how the brain actually performs these processes, it is not surprising that no instrument capable of replicating this process currently exists.

Since your beers are ultimately sold based on the consumer's response to their flavor, it is essential that you taste your beers before sending them out the door. In fact, if this book could be distilled into one simple statement, it is that you need to make sure you are tasting every batch of beer you make.

Sensory Quality Control

While humans have informally used sensory to assess quality for hundreds, if not thousands, of years, rigorous application of scientific principles to sensory work in quality control is a comparatively young discipline. Historically, sensory evaluation of consumable products fell to an individual who would make all flavor-related decisions based on their expertise (Lawless and Heymann 2010, 4). A certain mystique or romanticism surrounds the notion of a single expert using their palate to guide the characteristics of a brand, and some companies do still make decisions in this manner.

However, dependence on a single person carries a distinct set of drawbacks. What happens when this person gets sick? Or has to travel? Or wants to take a vacation? Or leaves the company, retires, or dies? For companies that tether their financial fortunes to an individual, succession planning is often a significant headache when that person decides to move on. Look to the lambic producers of Belgium—most employ a single blender who carefully shapes the flavor of their beers, often spending decades of his or her career at a single brewery. Each retirement and subsequent transition carries a significant risk as fans of the brewery hold their breath, waiting to see if the new blender's beers can match the quality of the former's.

By the 1960s, producers of consumable products had begun to shift from reliance on a single expert to using a panel of trained tasters (Stone and Sidel 2004, 10). Today, the benefits of using a panel seem obvious. In addition to mitigating the concerns relating to the absence (whether temporary or permanent) of a single expert, a panel produces more reliable assessments of products and, consequently, more consistent results. There is no such thing as a perfect palate—everyone has strengths and weaknesses largely determined by their biological makeup, and most people are anosmic to (i.e., unable to smell) at least a few flavor compounds. Using a panel of tasters mitigates the risk of relying on a single taster who may be blind to an important flavor compound. Furthermore, an individual's sensory acuity can vary from day to day based on a variety of factors, including how much sleep they got the night before, what they have eaten that day, and even their general mood (Meilgaard, Civille, and Carr 2016, 49–50). If all decisions are made based on one person's palate, the potential for variation is immense. Once again, using a panel of tasters helps reduce errors arising from variations in a single taster's results.

While sensory work is an essential complement to instrumental methods of quality control in beer production, using human subjects introduces a number of issues that must be considered. In sensory work, your panelists are your instruments, and just like mechanical instruments, they require some calibration and upkeep to ensure they deliver usable data. However, humans are not mechanical instruments. I bet you have never worried about your refractometer yielding bad measurements because it is having a bad day. You do not need to worry about your pH meter unexpectedly giving an incorrect reading due to confirmation bias. And you certainly do not need to worry about your microscope just not showing up due to a lack of motivation. Yet, with human panelists, these possibilities are all on the table. To some extent, you will always have to contend with variability, bias, and motivation issues when working with human panelists, but the data they will provide makes it all worthwhile. We will address these common concerns throughout the book in an attempt to limit their impact on your data.

BEYOND QUALITY CONTROL

While your sensory program should initially focus on quality control, you will quickly find a number of other novel and exciting uses for sensory evaluation within your brewery. Once you have established protocols for product release, you can begin to use your panelists for a wider group of tasks. You can use your panel to help determine the acceptable shelf life of your products. You can also work with your panel to evaluate the sensory properties of your ingredients. This can help you better understand the beers you are making and will aid you in any recipe reformulation or substitutions that you need to make. This information is also invaluable when constructing new recipes or designing new brands. You can also use your panel to effectively prototype and refine new beers rather than just having the head brewer make all of the decisions on their own. This greatly increases the likelihood that the beer will succeed when you take it to market. These topics are covered in depth in chapters 11–13.

BUILDING A CULTURE OF SENSORY

Most breweries consider quality or, more directly, the production of high-quality beer, a core value. However, quality is not solely in the hands of a brewery's production staff. Yes, their actions and decisions do largely determine whether quality beer is produced, but broader quality control work helps verify that quality beer is being made, while alerting production if quality falters. While only a small proportion of employees at a brewery may actually make the beer, quality should be important to everyone. Building a sensory program by soliciting panelists from different departments within the brewery will give a wider group of people a stake in the pursuit of quality. To ensure that employees throughout the brewery can freely participate, you should first secure buy-in from brewery leadership.

WHO WILL LEAD THE CHARGE?

If you are trying to establish a new sensory program, you should begin by determining who will serve as panel leader. Small organizations will likely not have the resources to dedicate an employee's entire focus to sensory work, but, in all honesty, that level of time investment is not required at this stage. A bare-bones sensory program can yield valuable insights from structured tastings conducted on a weekly basis, which require only a couple of hours of the panel leader's time each week. With that in mind, I would argue that the single most important criterion for selecting a panel leader is to choose someone with a natural interest in sensory evaluation. While you might want to place a sensory expert in that role, you may not currently have a sensory expert on staff. However, if you select someone with an inclination for the subject matter, they will enthusiastically dive into the task of building your program and can develop expertise along the way. And, given the fact that you have picked up a book on how to build a sensory program, that person might very well be you!

For your sensory program to succeed, the heads of your organization must believe in its importance. Their buy-in should be emphatically communicated and demonstrated to the rest of the staff to make sure that all employees understand the purpose and value of the program. In small organizations, this does not usually present much of an issue. With fewer employees, most people at small breweries wear multiple hats; departments are not necessarily distinct from one another, which facilitates communication between key stakeholders. In the smallest operations the majority of employees may even sit on the sensory panel.

As breweries grow, communication becomes both more difficult and more essential. To make use of the best palates at your brewery, you may pull panelists from a variety of different departments. Since panel sessions necessarily take time away from an employee's core duties, department managers need to understand that sensory work is both a valid and valuable use of their employees' time. Some people, particularly those who have never sat through a panel session, may be under the impression that sessions merely offer panelists a break from work to have a couple of beers. This could not be further from the truth. Panel work requires tasters to approach beer from an analytical perspective—a challenging exercise that demands intense focus. Clearly communicating decisions informed by sensory data can help demonstrate the value of your program to any potential detractors.

Bear in mind that it is not enough for managers to simply buy into the importance of your sensory program. To truly build a company-wide culture of sensory, managers should attempt to convey its importance to any of their direct employees that serve as panelists so that they feel encouraged to continue participating. Similarly, occasional managerial participation can help drive enthusiasm for the program. Often, sensory data will be aggregated prior to evaluation, so the results from a high-ranking member of the brewery carry no more weight than those produced by a taproom barback. But if employees see the brewmaster or company head sitting down to taste at a panel session, the importance of the program becomes unquestionable.

BEST PRACTICES AND PRINCIPLES FOR RUNNING A SENSORY PROGRAM

While conducting research to develop this book, I engaged leaders of sensory programs from small and large breweries across the country to learn how their programs operate. Each program was unique, with specific elements dictated both by the culture and structure of the brewery. However, the best programs shared certain common elements, a set of unspoken principles that shaped the way sensory techniques were used to improve the beers produced at the breweries in question. These principles will come up time and time again throughout this book because they are the bedrock of an intelligently designed sensory program. Given the vast number and variety of breweries around the world today, it is impossible to offer a one-size-fits-all approach to implementing a sensory program. Rather than giving prescriptive advice, my hope is that you can use the five principles that follow to tailor your program to meet the specific needs and unique considerations of your brewery.

Each sensory test you perform should be designed to answer a specific question. That question can take many different forms. Is this IPA true to its brand specifications and fit for release? Is our American Stout still salable five months after packaging? Does swapping Pilsner malt A for Pilsner malt B change the flavor of our Helles? Regardless of what type of test you are using, you should be able to clearly identify why you are running that test and what you hope to learn from it. You should never find yourself performing a test "just because." This is the quickest way to waste your panelists' time, resources, and energy. Every test should serve a purpose.

Each sensory test should produce actionable data. As a corollary to the previous point, your tests should also produce data that allow you to make decisions. The decision could be as simple as releasing a beer to the trade versus holding it back for further testing. The "action" could involve maintaining the status quo, for example passing on a new ingredient supplier in favor of continuing with your current one. The important point is that the questions you are asking and the tests you are performing generate data that inform decisions. You should avoid collecting data for the sole purpose of having more information.

Each sensory test should have action standards in place. Action standards describe the actions taken depending on the result of a test. For example, in the context of a true-to-target test used for product release, the action standard for a sample that passes might be to release the beer to the trade, while the action standard for a sample that fails might be to hold that batch of beer for further testing. The exact action standards will depend on the question that you are trying to answer and the test being performed.

You should always strive to establish action standards in advance of performing a test. If action standards are not in place, it can be more challenging to make the right decision in the face of a negative result. Ideally, action standards are developed in a logical way to help expedite decision-making in difficult situations, rather than having to determine what to do on the fly in the emotionally charged moment when a sample fails a product release panel. Without action standards in place, the temptation may be to downplay the significance of the results of a given test. Furthermore, designing action standards for your tests will help identify any tests where you are simply monitoring a process without any real purpose. If different results on a given test do not trigger different actions, you should examine carefully whether that test is worth performing at all.

Favor rapid tests that determine whether additional testing is necessary. To make the best use of your panelists' time, endeavor to use tests that are simple and can quickly and accurately flag potential issues. The true-to-target test (see chap. 6) is a prime example of a test that will rapidly distinguish between samples that conform to specifications and samples that fall outside the acceptable range of variation. We will explore a number of different applications for the true-to-target test in chapters 10–12. The implications of this principle will become clearer after covering the different types of sensory tests presented in chapter 7.

Do not allow your sensory program to exist in a silo. To ensure that your sensory program is considered relevant throughout the brewery, you must make sure that the sensory department communicates effectively with other departments so that other departments can use sensory data to make decisions. Ideally, data generated by your sensory program will impact and improve operations throughout the brewery, and this can only occur with strong communication.

DEVELOPING YOUR OWN SENSORY PHILOSOPHY

Given the varied landscape of different breweries in existence today, I cannot offer a single specific model for what an ideal sensory program might look like at your brewery. However, I can definitely state that sensory techniques can be used to improve the quality of your beer. This statement is true whether you are a large brewery that only produces a handful of core brands or a nanobrewery that has never brewed the same beer twice. Use the principles above and the information presented in the following chapters to shape your approach to building and running a sensory program. With time and experience, you will determine what is most relevant for your program and what works best for your brewery. The specific decisions you make to shape your program are in your hands. Good luck!

2
THE SENSES:
HOW FLAVOR IS MADE

The human experience is built upon a scaffold of sensory perceptions, themselves the product of signals produced by our sensory neurons. The nexus of this activity—the brain—filters this never-ending cascade of information and transforms it into a cohesive perception of our environment. Thus, our core senses—sight, sound, smell, taste, and touch—define our individual realities.

Our brains are exceptionally powerful processing machines, capable of culling tremendous amounts of data to leave us with only the most essential pieces necessary to understand what is happening in our immediate environment. Because the brain has to accomplish this task both rapidly and constantly, it uses pattern recognition to avoid analyzing each individual piece of sensory information (Shepherd 2012, 83). Common examples include facial recognition and language skills. We do not need to carefully examine the shape of someone's nose or the color of their eyes to ascertain their identity. Using the composite of their features combined with memory, we are able to call to mind a name. With language, we do not need to analyze the individual phonemes that make up a given word, we come to instantly recognize the meaning implied by certain groupings of sounds. In a similar way, flavor recognition functions through pattern recognition—the brain's processing of certain sensory inputs combined with experiential memory yields a perceptual experience.

Unlike our use of sight to recognize a face or our use of sound to process language, we do not have a single sense that governs flavor perception. Many

texts rightfully state that our sense of smell dictates much of our flavor experience, assisted by our sense of taste, but this provides an incomplete picture of flavor. Flavor is perhaps the ultimate example of our brain's ability to synthesize sensory inputs from different sources, in this case using all available senses to produce a single, unified perception. These patterns form robust, interconnected neural networks such that a single sensory input can produce an entire flavor image from our memory. A waft of freshly baked baguette can elicit the image of golden-brown bread crust, the toothsome crunch of the crust giving way to the pillowy interior, and even the crackle of that crust as you press it between your teeth. When eating a baguette, our brain combines all of these sensory inputs into a singular flavor experience. After having the experience, any one of these inputs can serve to trigger the entire flavor memory.

When we perform sensory analysis on a beverage, we fight against our brain's natural impulse to aggregate data into a single impression. Instead, we seek to break apart that unified perception into its component parts to make it easier to evaluate. It is challenging to pull apart a unified flavor, but if we can identify the component cues for taste, aroma, or appearance, it becomes easier to accurately assess and describe the pieces that make up beer flavor. By better understanding how each of our senses operates to produce a given flavor experience, we can better understand how to leverage each sense to yield accurate sensory data.

THE SENSE OF TASTE

The sense of taste evolved as a gatekeeper to the rest of the body, serving animals in their pursuit of sustenance. Different tastes offer dramatically different messages. Sweetness promises energy and nutrition, whereas high levels of acidity might signal spoilage. Bitterness in particular warns of danger; many poisonous plants contain bitter-tasting compounds. This evolutionary fact is borne out in animals today, as carnivores typically have fewer taste buds compared to herbivores (Jiang et al. 2012, 4956). Early humans similarly used their sense of taste to inform what they should and should not eat. Our present predilections for certain tastes are driven by the same factors that guided our biological ancestors through a dangerous and uncertain gustatory landscape.

Anatomy of the Tongue

The tongue is both a physical organ and a sensory organ. In the physical sense, we actively use our tongues to move food within our mouths. However, in the context of flavor perception we are primarily concerned with the sensory functions of the tongue.

Our tongues are covered with small, elevated structures called papillae, of which there are four types: filiform, fungiform, foliate, and circumvallate. Filiform papillae, the small, whitish-pink bumps that cover the majority of our tongue, are the most prevalent but also the least consequential to taste perception because they do not contain taste buds. The other three types of papillae all contain taste buds and differ primarily based on their structure and location on the tongue. The fungiform papillae appear as red, club-shaped bumps on the tip and surface of the tongue and have taste buds on their upper surface. The foliate papillae are found along the rear sides of the tongue as a series of short vertical folds and bear many taste buds. Lastly, the circumvallate papillae form a V-shaped row near the back of the tongue; each circumvallate papilla is dome-shaped, with taste buds along its upper sides (Lawless and Heymann 2010, 29).

While a taste scientist may find value in understanding the different types of papillae, we are primarily concerned with taste buds and their function. Depending on various factors, including genetics and age, humans will have between 2,000 and 8,000

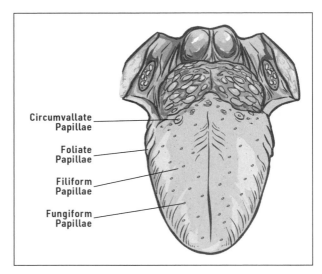

Figure 2.1. Diagram of the tongue, including the locations of the different types of papillae. © Getty/corbac40

taste buds at any given time, with each individual taste bud containing anywhere from 50 to 150 taste receptor cells.[1] When we eat or drink, our saliva serves as a solvent, carrying taste molecules into our taste buds to interact with the taste receptors.

A pervasive and insidious myth that exists in the realm of flavor and taste knowledge is that of the well-defined tongue map. Many of you have probably seen some iteration of the map; it is a common enough feature of most pop-science literature discussing taste. This map, however, has long been known to be wrong (see "The Tongue Map Myth"). The idea that the tongue is divided into clearly delineated regions in which we perceive each of the tastes has been thoroughly debunked over the past several decades (Collings 1974, 169).

Basic Tastes

Compared with our other senses, taste features a rather small set of distinct perceptions. While the different taste sensations do cover a wide range of compounds, science currently defines only six categories of taste: sweet, bitter, sour, salty, umami, and fat. However, other sensations, such as capsaicin heat (i.e., spiciness), seem like good candidates for basic tastes. So why do capsaicin and other, similar characteristics not qualify? It turns out that in order for a class of compounds to be considered a unique, basic taste, it must meet five criteria (Keast and Costanzo 2015, 2):

---

1   Elizabeth Bernays and Reginald Chapman, *Encyclopaedia Britannica Online*, s.v. "Taste bud," accessed September 8, 2019, http://www.britannica.com/science/taste-bud.

THE TONGUE MAP MYTH

The tongue map packages a few distinct errors made several decades apart into the widely distributed diagram that many know today. The first was a study conducted in 1901 by German scientist David P. Hänig, intended to map out variations in taste sensitivity across the different regions of the tongue. However, the study was poorly designed, featuring too small of a sample pool to produce relevant results. Furthermore, the results Hänig obtained indicated only slight differences in sensitivity across the tongue, rather than a series of regions that could each perceive only one taste. However, the real damage was done when American psychologist Edwin Boring included Hänig's findings in his seminal work, *Sensation and Perceptions in the History of Experimental Psychology*, published in 1942. In an effort to illustrate Hänig's results, Boring created a unitless graph that removed any nuance present in the original report, greatly amplifying the slight sensitivity differences that Hänig encountered.

Boring's graph led to further misinterpretation, resulting first in a tongue labeled with the tastes assigned to different regions, followed by the addition of boundaries cordoning off the tastes into their own distinct realms, yielding the misleading tongue map we know today (McQuaid 2015, 2–5). Numerous studies have been performed that disprove the ubiquitous tongue map, and yet it persists in our collective consciousness. Why the map boasts such staying power is beyond my knowledge, but it still appears in texts on wine and spirits, as well as in children's science textbooks. My best guess is that its inherent simplicity and our natural desire to understand the way our senses work make it a comfortable tool to grasp, and a difficult one to let go of.

1. It must be a distinct class of affective stimuli.
2. There should be a transduction mechanism that includes receptors to change the chemical code of the stimulus to an electrical signal.
3. There must be neurotransmission of the electrical signal to processing regions in the brain.
4. There should be perceptual independence from other taste qualities.
5. There must be physiological effects after activation of taste bud cells.

That last point is especially important in this case and explains why spiciness does not make the cut. Logically enough, only molecules that interact with taste receptors can qualify as basic tastes. As we shall see in the section on mouthfeel, capsaicin, while primarily active within the mouth, targets an entirely different sensory system.

Sweet

Sweet taste evolved as a mechanism to detect nutritive value in food, as simple sugars represent a convenient source of energy for most animals. This taste served our ancestors well when energy sources were in short supply, but in our current sugar-saturated culture the brain's fondness for sweet tastes can have unfortunate consequences.

Sweet taste receptors are one of three categories of taste receptors that are part of the G protein–coupled receptor family of proteins, the others being umami and bitterness. A number of different simple sugars can activate sweet taste receptors. These simple sugars include glucose, sucrose (table sugar), fructose (fruit sugar), lactose (milk sugar), and maltose (malt sugar), with different sugars producing differing levels of sweetness (table 2.1). The artificial sweeteners saccharin, aspartame, and sucralose also stimulate sweetness receptors, and in some cases are hundreds of times more effective at producing a sweet response than natural sugars (Joesten, Castellion, and Hogg 2007, 359). This allows food and beverage manufacturers to substitute sucrose or other sugars with artificial sweeteners at very low levels while still achieving a similar level of perceived sweetness.

In beer, sweetness results from residual sugars remaining in the beer after fermentation. Most beers have relatively low residual sweetness following fermentation, although styles with lower attenuation can contain more significant levels of sweetness, and beers that employ lactose will also display some amount of sweetness. In addition to the actual level of sugar present, several other beer characteristics can influence the level of sweetness we ultimately perceive. Both

G PROTEIN–COUPLED RECEPTORS

G protein–coupled receptors (GPCRs) are a large family of receptor proteins that exist embedded in cell membranes. Different GPCRs serve to detect specific molecules outside of the cell by binding to them (e.g., sweet taste GPCRs bind with sugars like glucose or maltose). Upon this binding interaction, the GPCR initiates a signal transduction pathway that triggers a cellular response. GPCRs are involved in signal transduction in a number of different systems, including taste, aroma, and sight.

Different genes encode for different types of GPCRs. Within the realm of taste, there are two key families of GPCR genes: the *TAS1R* group, which codes for sweet and umami receptors, and the *TAS2R* group, which codes for bitterness receptors.

Table 2.1 Perceived sweetness of different sweeteners relative to sucrose

Sweetener	Relative sweetness
Lactose (milk sugar)	0.16
Maltose (malt sugar)	0.33
Glucose (dextrose)	0.74
Sucrose (table sugar)	1.00
Fructose (fruit sugar)	1.17
Aspartame	180
Saccharin	300
Sucralose	600

Source: Joesten, Castellion, and Hogg (2007, 359).

bitter and sour tastes have a suppressive effect on sweet taste (Lawless and Heymann 2010, 31), so beers high in either of these attributes will display diminished perceivable sweetness relative to the amount of sugar present. Similarly, high carbonation suppresses the perception of sweetness (Hewson et al. 2009, 94), which contributes to the dry (i.e., not sweet) character of many highly carbonated Belgian styles.

Bitter

Counter to sweetness, bitter taste evolved as a way for our bodies to detect things that might harm us. An aversive taste, we know from birth to avoid bitter substances; several experiments have shown that babies will scrunch up their faces in disgust when presented with a bitter-tasting substance (Shepherd 2012, 124). However, as we age, we can learn to appreciate or even enjoy bitter things, a demonstration of how our hedonic perspective—our likes and dislikes—can change with exposure.

Bitter taste is also mediated by G protein–coupled receptors, but the bitterness receptor gene family is significantly larger than the one for sweetness and umami taste receptors. The human *TAS1R* gene family, which encodes the receptors for umami and sweetness, contains just three genes. The *TAS2R* family, which encodes the receptors for bitterness, includes 43 currently identified genes (Bachmanov and Beauchamp 2007, 396).

One interesting feature of bitter taste in beer is that it tends to increase on the palate after exposure, rather than producing an initially strong sensation. Upon tasting a beer with a moderate to high level of bitterness, tasters will often find that the level of perceived bitterness peaks 15 to 30 seconds after swallowing rather than on initial exposure.

Today, many different bitter ingredients have found their way into modern foodstuffs, including green vegetables such as broccoli, Brussels sprouts, and kale, and also chocolate, coffee, hops in beer, cinchona bark in tonic water, and gentian root and other bittering substances used in the production of bitter liqueurs. Each of these ingredients contains some sort of bitter-tasting molecule, such as quinine in cinchona bark, caffeine in coffee, and various alpha acids and iso-alpha acids in hops. Over 500 different compounds have been identified that elicit a bitter taste response in humans (Wiener et al. 2012, 413).

In beer, we are primarily concerned with alpha acids and iso-alpha acids, referred to as humulones and isohumulones, respectively. Brewers use the international bitterness unit (IBU) to measure the bitterness level of their beers, with 1 IBU corresponding to 1 mg/L dissolved iso-alpha acid in the finished beer.[2] The IBU level can vary widely between different beers at the discretion of the brewer, who controls the level by adjusting the amount of hops used and the timing of the hop

---

2   One milligram per liter (mg/L) is equal to one part per million (ppm); 1 mg/L = 1 ppm.

additions during the brewing process. However, as with sweetness in beer, bitterness perception operates within the complex matrix of beer, so the level of bitterness perceived often does not correlate directly to the IBU number of a given beer. Because sweetness suppresses bitterness, residual sweetness can dampen the perceived bitterness of a beer. Furthermore, carbonation's suppressive effect on sweetness does not extend to bitterness. As a result, higher levels of carbonation can elevate the perceived bitterness of a beer by suppressing any balancing sweetness present.

While bitterness plays an important role in many global cuisines, it is ultimately an aversive taste and there remains a large segment of the population that prefers lower levels of bitterness. This can be seen in recent trends relating to IPA-style beers. While highly bitter IPAs achieved relatively widespread popularity over the past decade, New England IPAs now appear to enjoy even broader acceptance. New England IPAs often feature intense hop flavors of tropical fruit coupled with lower perceived bitterness and higher perceived sweetness. The current craze surrounding this style may indicate that while many people enjoy the aroma of hops, a significant segment of individuals found the high levels of bitterness in traditional IPAs objectionable.

Sour

Sour taste receptors detect acids present in foods and beverages. Unlike the tastes of sweetness and bitterness, sour taste receptor cells do not have receptors on the outside of their cells. Instead, they primarily measure dissolved hydrogen ions in solution, much as a pH meter does. Sour taste receptor cells have hydrogen ion channels embedded in their cell membranes, which allow hydrogen ions (protons) to enter the cell and trigger a response. Additionally, there is some evidence that these cells can also take in weaker acids and interact with them internally (Ye et al. 2015, 229).

Acidity is a common element in many foodstuffs and is naturally abundant in most fruits. Additionally, fermented products often contain some amount of acid, and particularly acidic products (e.g., yogurt and sauerkraut) are fermented with lactic acid–producing bacteria, such as *Lactobacillus* or *Pediococcus* species. The most common organic acids that elicit a response from acid taste receptors include lactic acid, acetic acid, malic acid, citric acid, tartaric acid, and

SUPERTASTERS

The word "supertaster" gets thrown around a lot in the world of beer tasting. The idea of someone with exceptional tasting ability, capable of detecting tiny nuances in each beer they evaluate, plays strongly to our collective imagination. However, that is not actually what the term supertaster describes. Supertasters have a higher-than-normal concentration of taste buds on their tongues, which causes them to experience tastes more intensely, particularly bitter and acidic tastes. But this does not make supertasters inherently better at tasting than non-supertasters. In fact, this often leaves the supertaster so overly sensitive that they end up being a picky eater, avoiding certain tastes due to the unpleasant sensations they produce. However, assuming that their heightened sensitivity does not lead them to avoid beer, supertasters can still perform well in a panel setting. Your panelists will naturally exhibit variation in their sensitivities to different tastes and aromas. The entire principle behind using a panel of tasters relies on the fact that the group functions better than a single taster would on their own, because the individual strengths and weaknesses of each panelist balance one another.

oxalic acid. The most common acids found in beer include lactic and acetic acid, though other acids such as malic, citric, and tartaric acid may appear, particularly in cases where fruits are used. Some acids, such as acetic acid, also exhibit aromatic activity and have distinct flavors associated with them, but most of these acids simply vary in perceived intensity.

Beer is acidic, and all beer contains some amount of acid that serves to balance the sweetness and other elements of the beverage. What we think of as sour beers typically carry significant levels of lactic and other organic acids, usually either from a bacterial fermentation, kettle souring (a rapid bacterial fermentation), or direct addition. The addition of fruit or other ingredients with high levels of acids can also contribute high levels of acidity to beer.

Furthermore, when carbon dioxide dissolves in a liquid it forms carbonic acid, further increasing the acidity and contributing to sour taste (Hewson et al. 2009, 94). Carbonation impacts many aspects of beer, affecting mouthfeel, suppressing sweet taste, and contributing to sour taste. Although carbonation alone will not make a beer taste sour outright, beers with high levels of carbonation can show slightly increased perceivable sourness.

Brewers use two different measurements to track the acidity of beer: pH and titratable acidity (TA). While pH readings offer significant utility when monitoring brewing processes like mashing and fermentation, they do not correlate well with the perceived sourness of a finished beer. Some producers of sour beer feel that TA better approximates the perceived sour taste of a beer, but this correlation does not always hold up either. Just like with IBUs and perceived bitterness, the perceived sourness of a beer is not governed solely by the acids present. Sweetness can greatly diminish perceived sourness—just think of the effect of adding sugar to lemon juice to make lemonade. Both pH and TA fail to account for other beer characteristics like residual sugar or alcohol levels, and so cannot replace sensory evaluation of sourness.

Several other organic acids present in beer do not have a large impact on the level of sour taste but can have a tremendous impact on the overall flavor profile of the beer. For example, isovaleric acid and butyric acid are odor active at extremely low levels and produce unpleasant aromas when found in beer. Because of their high level of odor activity, they will affect the aroma profile of a beer long before they impart any notable acidity. In order to derive sour taste from either of these compounds they would need to be present at levels several thousand times higher than their odor thresholds—a truly unpleasant thought!

Both sourness and bitterness can become unpleasant at high levels and can induce a stinging or pain-like reaction in tasters. As a consequence, some tasters experience sour-bitter confusion, in which they confuse sour tastes for bitter tastes or vice versa. One key anecdotal difference between the two is that sour taste tends to present a stronger upfront attack and lingers less, whereas bitterness tends to linger on the palate, typically peaking in perceived intensity 15 to 30 seconds after the bitter stimulus is swallowed. Training with sour and bitter solutions can help tasters to distinguish between these two tastes (see p. 49).

TART VERSUS SOUR

Tasters use a variety of different terms to describe sour taste in beer, "sour," "tart," and "acidic" being three of the most common. I have been asked in numerous tasting sessions over the years how to differentiate between these terms. However, the terms really just name the same taste sensation, and any difference between them typically stems from the way an individual decides to describe the intensity level of a sour taste. For example, in most cases, tart would be used to describe a lower degree of acidity than sour. Even this usage is not universal though. Within your panel, my recommendation would be to pick one of these words and then modify it with "low," "medium," and "high." You should use every opportunity you have to make descriptive vocabulary among your panelists as uniform as possible.

Salty

Salt taste receptors have the simplest mechanism of the tastes, working by directly detecting sodium ions, in addition to other alkali metal ions. Salt taste receptor cells have sodium ion channels embedded in their cell membranes, allowing sodium ions to enter the cell and elicit a taste response. Although sodium ions are the most potent activator of this response, alkali metal ions of similar size, such as potassium ions and lithium ions, can also activate the salt taste response, albeit to a lesser extent.

While salt levels play a key role in taste responses to different foods, most beers do not contain appreciable levels of sodium chloride (table salt). A notable exception is Gose, a sour German wheat beer brewed with sodium chloride. When added at low levels, sodium chloride will not stimulate an outright salt taste response, but it will boost the perceived sweetness of the beer (probably by suppressing bitterness) and increase the perceived body of the beer. At higher levels, the saltiness becomes perceptible, with extreme levels leading to an unpleasant impression of saltwater.

Umami

Umami has long been known in East Asian cultures, having first been isolated and proposed as a basic taste in Tokyo in the early 1900s (McQuaid 2015, 234). Western science was a bit slower to adopt it as the fifth basic taste, so understanding of umami in Western cultures still lags behind. Further hindering its penetration into common vernacular, the nature of umami taste is more challenging to define than the other more familiar tastes. The direct translation of *umami* from Japanese is something along the lines of "pleasant savory taste" or "deliciousness," which does not help much in the way of specificity.

Umami detection operates through G protein–coupled receptors, like sweetness and bitterness, and the umami receptor is coded for by two genes of the *TAS1R* family. The receptor primarily detects levels of the amino acid glutamate in addition to some ribonucleotides like guanosine monophosphate (GMP) and inosine monophosphate (IMP). Most discussion of umami taste revolves around monosodium glutamate (MSG), which has been maligned as an additive in food even though it appears naturally in many foodstuffs and is not inherently harmful (Obayashi and Nagamura 2016, 6).

Since the taste of umami defies simple description, many tasters turn to foods high in umami when learning to recognize its characteristics. Foods with elevated levels of umami include most protein sources, such as meat, fish, shellfish, ripe tomatoes, and mushrooms. It also appears at significant levels in many fermented foods, such as soy sauce, fish sauce, cheese, and yeast extracts such as Vegemite and Marmite.

Although umami rarely appears in beer, significant levels in yeast extracts give us a clue as to the one potential source of umami taste in beer. Yeast autolysis can lead to the development of umami taste in beer. At low levels in aged beers this can add a positive complement, but even a moderate level of umami in beer is considered undesirable.

Fat

Recent research has demonstrated evidence for a taste receptor that responds to fat molecules, specifically certain long-chain fatty acids. The sensation associated with fat taste is not a pleasant one. The specific molecules that produce a response yield a highly unpleasant, rancid taste experience. As such,

AUTOLYSIS

Autolysis describes the self-destruction process that yeast cells undergo upon cell death, in which their cells break open (lyse), spilling the contents of the cells into the beer. Autolysis can lead to a number of less-than-pleasant flavors. In addition to umami due to the release of glutamates, aroma compounds such as mercaptan or methanethiol (commonly described as smelling like powdery-rind cheese or hot garbage, literally) can also accumulate in the beer. Tasting yeast extracts like Vegemite or Marmite can give you a good idea of the sensory impact of autolyzed yeast, as autolyzed yeast is the primary ingredient in these products.

Autolysis flavors can appear in very old bottle-conditioned beer as the yeast contained in the beer gradually dies. However, autolysis flavors can also show up in younger beers when the yeast has been subjected to excessive stress and undergoes cell death earlier than expected. Yeast stored too long or too warm in between pitches can lead to significant levels of autolyzed yeast in the pitch before the yeast is even used. Also, leaving yeast overly long in the cone of cylindroconical vessels following flocculation can also increase rates of autolysis due to the tremendous hydrostatic pressure placed on the yeast at the bottom of the tank. This effect becomes more pronounced as tanks get larger and taller, increasing the necessity for proper yeast management in larger breweries.

most commonly consumed foods—even foods high in fats—stimulate little to no response from these fat taste receptors. These receptors may have evolved to serve a similar purpose to other aversive tastes (e.g., bitterness and sourness) to help our primitive ancestors avoid spoiled meats and other rancid fatty foods that might induce sickness or disease.

The stimuli and mechanisms of fat taste are still being studied, but many taste scientists consider fat to be a credible sixth taste (Keast and Costanzo 2015, 6). For all intents and purposes, this taste does not appear in beer, so panelists do not need to be trained to detect it.

Other Elements Influencing Taste Perception

Adaptation

One of the challenges with panel tasting is that panelists will rarely (if ever) interact with only one sample. With multiple samples in play, you must consider the role that adaptation of each sense plays on the perceptions recorded by your panelists. Taste does show adaptive effects in that, when exposed to a certain taste, panelists become temporarily desensitized to that taste (Lawless and Heymann 2010, 30–31). If, for example, a panelist tastes a beer with a high level of bitterness they will adapt to that level of bitterness. If they then move directly to another sample their perception of the level of bitterness present in that second sample will be skewed by their adaptation to the bitter taste of the first. To combat adaptation, make panelists aware of its impact and encourage them to use water to cleanse their palates between samples. Time permitting, you can also suggest that panelists take a short break (say, 30 to 60 seconds) between each sample to help mitigate the effects of adaptation.

Taste Interactions

When present in combination, the different basic tastes exhibit predictable interactions. The four classic tastes—sweet, sour, salty, and bitter—tend to suppress one another to differing degrees. Food and beverages rarely display just one taste modality, but rather exist as an intricate balance of the different tastes. Fruit juices, for example, balance sweetness and acidity, with the sugar present diminishing the perception of the acid and the acid similarly diminishing the perception of the sugar. Many pharmaceutical compounds have unpleasant levels of bitterness, so pharmaceutical companies will often add salts to their preparations to help reduce the perceived bitterness (Keast, Breslin, and Beauchamp 2001, 441). Beer demonstrates an intricate balance between sweetness, bitterness, and acidity. Anecdotally, many brewers know that increasing bitterness can reduce the perception of sweetness in a beer. And one need only look at the history of Berliner weisse, an aggressively tart beer, to find patrons often adding fruited syrup to help blunt the perception of acidity.

Of the different tastes, sweet substances tend to most strongly suppress other tastes and tend to be the least suppressed themselves. Furthermore, interactions between sweetness and saltiness do not appear to be entirely suppressive. Yes, sweetness will suppress salty taste, but at low levels saltiness actually appears to enhance sweet taste, one of few enhancing interactions reported between tastes in the literature. Furthermore, in a complex mixture like beer that exists as a delicate balance of sweetness and bitterness, the addition of salt will significantly depress the perception of bitterness, another reason why low levels of salt in beer can enhance sweet taste (Keast, Breslin, and Beauchamp 2001, 441).

Temperature Effects on Taste and Aroma

While temperature plays an important role in taste perception, its impacts are complex and not fully understood. You may have heard that cold temperatures "numb your taste buds" and that this is why tastes tend to get more intense as temperatures increase toward room temperature. However, the mechanisms are significantly more complex than a simple numbing effect, not only varying from taste to taste but from compound to compound within tastes (Lipscomb, Rieck, and Dawson 2012, 4–6). For example, temperature increase is found to increase the sweetness of sucrose, but temperature increase has no impact on the sweetness of saccharin (Talavera et al. 2007, 378). The situation becomes even more complicated as other tastes are added to the mix. As a result, the way the taste of a specific beer may change as temperature increases is difficult, if not entirely impossible, to predict.

The mechanism by which temperature affects aroma is a bit more straightforward, driven as it is by thermodynamics rather than physiological systems within your body. As temperature increases, the volatility (i.e., the tendency of a compound to vaporize and so become airborne) of aromatic compounds increases, so serving a beer at higher temperatures will drive higher levels of overall aroma. However, the exact impacts on the overall perceived aroma of the beer are, once again, nearly impossible to predict, because increased temperature will not affect all aroma compounds equally. While all compounds will become more volatile as temperature increases, they will not necessarily do so at the same rate. This means that a compound that is barely perceptible at 4°C (40°F) may greatly increase in volatility as the temperature rises to

13°C (55°F), going from a background note to a dominant feature of the beer's profile.

Serving your beer at different temperatures will lead to different flavor experiences in an unpredictable fashion. To mitigate these effects, you must take temperature into account when serving samples to panelists, and you should endeavor to serve samples at a consistent temperature, both across a single session and from session to session.

FLAVOR TRIPPING

One of the strangest peculiarities of taste involves so-called "miracle berries", the curious fruit of the *Synsepalum dulcificum* plant. These berries contain a protein called miraculin. Miraculin has no taste of its own, but when consumed it binds to the tongue and causes acidic foods and beverages to trigger a strong sweet taste response (Theerasilp and Kurihara 1988, 11536). Saliva will eventually wash the miraculin from the tongue but the effects can persist for as long as an hour.

Eating the berries, sometimes referred to as "flavor tripping," can lead to a number of novel tasting experiences. Citrus fruits like lemons and limes, normally grimace-inducing if consumed on their own, can be eaten by the slice, triggering intense, candy-like sweetness. Vinegar comes across as sweet as syrup. The experience is certainly one of cognitive dissonance, as our brain tries to comprehend how these intense, normally aversive sour tastes could possibly trigger an entirely different sensation.

Thus far, there have not been a lot of practical applications for miraculin in food or beverages, and the US Food and Drug Administration currently has a ban on miraculin as a food additive. However, some chefs have experimented with designing food experiences around tasting various items following consumption of miraculin. Although currently an unexplored frontier, it might be interesting to see a brewer produce a beer intended to be consumed only after ingesting miraculin.

THE SENSE OF SMELL

Aroma, like taste, has long served a role in our pursuit of sustenance (Sarafoleanu et al. 2009, 196). Human aroma detection begins in a region known as the olfactory epithelium, which is located on the roof of the nasal cavity about 7 cm (2.8 in) up from the nostrils. The olfactory epithelium is made up of several different types of structural cells along with cells known as olfactory receptor neurons, which are sensory neurons containing the receptors that interact with different odor molecules. While the total number of distinct odor receptors is currently unknown, best estimates suggest that the mammalian genome includes as many as 1,000 different genes encoding for aroma receptors, with about 350 of these active within humans (Lawless and Heymann 2010, 34). However, humans can smell many more than 350 different odorants due to the way the olfactory system works.

The mechanism behind our sense of smell has only been understood since the early 2000s, and scientists are still learning how genetics affects the way individuals perceive aroma. In pursuit of this knowledge, scientists discarded a number of theories along the way, one of the most prevalent being some sort of lock-and-key mechanism in which our system had one specific receptor for each aroma molecule. Given the limited set of receptors we have and the thousands of different odorants that we can detect, the mechanism had to be more complex than a simple one-to-one match system.

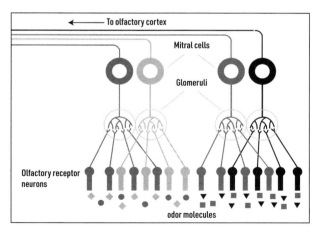

Figure 2.2. When odor molecules stimulate olfactory receptor neurons, the neurons pass the signal along to glomeruli in the olfactory bulb. This representation shows four receptor neurons connected to each glomerulus; in reality, each individual glomerulus receives input from thousands of identical olfactory receptors, effectively amplifying the signal sent by each receptor.

Olfactory receptor neurons transmit signals to specialized spherical structures within the olfactory bulb called glomeruli. Humans have around 1,000 glomeruli. Each glomerulus receives input from thousands of olfactory receptor neurons, all of which express the same type of receptor protein. In other words, each glomerulus is dedicated to receiving signals from a particular type of receptor. Thousands of the same type of olfactory receptor neuron converge to a single glomerulus, a mechanism that amplifies the signals sent by individual activated olfactory receptors. Breakthrough work demonstrated that a given odor molecule will stimulate multiple receptors to different degrees, and those activated receptors will produce a pattern within the olfactory bulb based on the glomeruli associated with those activated receptors (Shepherd 2012, 80). Since a single receptor can be activated by many different aromatic molecules and a single aromatic molecule can activate several different receptors, it creates the possibility for a nearly infinite number of potential combinations. This explains our ability to identify thousands of different odorants from a limited number of receptor types. Current thinking suggests that our brains recognize the different spatial patterns of activated glomeruli as discreet smells, much in the way we might recognize a pattern visually. This is one of the reasons why picking an aroma mixture apart molecule by molecule can be quite difficult—the brain tends to recognize the incoming aroma as a holistic image rather than a number of distinct features.

Aroma Thresholds

From a biological perspective, we define a threshold as the minimum amount of a stimulant required to cause a nerve cell to fire and subsequently deliver some sort of information to the brain. All senses exhibit this sort of threshold behavior. With sight, light must be of a certain intensity in order to elicit a response as something that we see. With sound, a certain decibel level is required before our brain registers a noise. However, both aroma and taste complicate this significantly, in that we have different thresholds for each compound. In the realm of aroma, the threshold levels of different compounds can differ significantly, by as many as several orders of magnitude. For example, to perceive acetic acid in beer it must be present at a concentration of at least 90 mg/L (90 ppm).[3] Compare that to a more odor-active compound like 3-methyl-2-butene-1-thiol (3MBT, the compound responsible for the skunk spray aroma in lightstruck beer), which we can smell at a mere 4 ng/l, or 4 ppt (De Keukeleire et al. 2008, 133).[4] This means 3MBT is more than ten million times more odor active than acetic acid!

With the mechanics behind aroma perception only recently identified, we still have a lot to learn about the way our sense of smell works. While we can use techniques from chemistry to examine the structure of a novel molecule and predict how it will react with other molecules, we do not have the ability to predict what it might smell like. In the same vein, we also do not know why different compounds elicit sensory responses at vastly different concentrations—we simply know it to be true through observation.

To further complicate matters, perception thresholds for individual compounds differ from one individual to the next. Differences in our individual genetics lead to differences in sensitivity to each and every aroma compound. Hence, the reported threshold for a given compound is the average of empirically determined thresholds across a large pool of individuals. Individual tasters will vary significantly in their sensitivity to specific compounds, but rarely will a given taster be weak across the board. By monitoring your panelists, you will develop a sense for which tasters are especially sensitive to certain flavors and which tasters are relatively insensitive to certain flavors.

Invariably, you will also find your panelists to be anosmic to certain flavors. Anosmia, technically defined, means that a panelist's detection threshold lies two standard deviations or more above the average detection threshold for a given compound. Put simply, if a panelist is anosmic to a specific compound they will effectively be blind to that flavor. Specific anosmias are entirely normal though—most panelists will likely exhibit specific anosmia to at least one or two different aroma compounds (Croy et al. 2016, 292). Furthermore, some compounds are notable

---

3   For 90 mg/L value, see "Acetic acid," under "Threshold," Aroxa (website), accessed September 29, 2019, https://www.aroxa.com/beer/beer-flavour-standard/acetic-acid. Even cited threshold values do not tell the whole story, as the perceivable threshold of a specific flavor compound may vary significantly depending on whether you are tasting a Pilsner or a barleywine!

4   One nanogram per liter (ng/L) is one-millionth the concentration of one milligram per liter. Thus, if 1 mg/L is equal to one part per million (ppm), then 1 ng/L is equal to one part per trillion (ppt).

THIS BEER SMELLS SWEET!

Although our brain builds a composite flavor experience using inputs from all of our senses, our senses themselves function independently of one another. Just like we do not have taste receptors for aromatic molecules like vanillin or diacetyl, we similarly do not have aroma receptors for sweet or bitter taste molecules. Yet I commonly hear untrained tasters describe the aroma profile of a given beer as sweet. If we cannot smell sugar molecules, what leads us to this misguided perception?

It turns out that the answer lies within a region of our brain called the orbitofrontal cortex (OFC). The OFC represents the first point within the brain where different sensory modalities come together, with some cells in the OFC responding to both taste and aroma inputs (Shepherd 2012, 115). Misidentified "sweet aromas" result from our brain's ability to recognize patterns of smells and tastes that typically occur together. Common examples in beer include fruity aromas of banana, pineapple, or figs, or so-called sweet aromatics like caramel, honey, marshmallow, or burnt sugar. When encountered in their natural forms, these foodstuffs are invariably sweet, as each of these foods contains high levels of sugar. Through repeated exposure, the brain forms an association between the signature aroma compounds of these foods and that sweet taste, eventually eliciting that sweet taste response simply through presentation of the aroma.

When we encounter these same aroma compounds in a beer, they are divorced from the sweetness inherent in their natural forms. You may encounter a beer that smells of prunes and dates and contains a high level of residual sugar, while another beer with a similar aroma profile may feature virtually no sweetness on the palate. The aroma of the beer offers no clue as to its level of sweetness. Furthermore, when you consume the beer, you will experience these aromatic elements retronasally, artificially elevating your assessment of the sweetness level regardless of how much sugar is actually present. Such aromas can confuse an untrained panelist, causing them to err in their judgment of the beer's taste profile.

With a bit of training and some clever techniques, we can outsmart our brains. When trying to accurately assess the taste profile of a beer, you can improve acuity by mechanically blocking aroma signals, pinching your nose like in the jellybean test (see pp. 23–24). In formal sensory evaluation, we strive for precision in assessment of the sensory experience. These sorts of simple techniques can help your panelists achieve their full potential.

for exhibiting significant rates of specific anosmia. Isovaleric acid, a compound found in aged hops (see p. 63), has an anosmia rate of 6% (Zhang and Firestein 2007, 1). Isobutyraldehyde, a key component of grainy malt aromas, rates even higher, at 36% anosmia in the general public (Moore, Forrester, and Pelosi 1976, 17).

Just because a panelist has a specific anosmia to a single compound does not mean that they cannot perform as a superb panelist. Obviously, you would not want to use a panelist anosmic to diacetyl for a diacetyl-specific sensory test like the diacetyl force test (see p. 62), but specific anosmias will not prevent panelists from performing well in true-to-target or descriptive panels. Rather, you should get to know the strengths and weaknesses of your panelists, including their anosmias, as this will help you better interpret their results and use your panel to its fullest extent.

Orthonasal and Retronasal Detection

Aroma molecules can travel through two distinct pathways to reach the olfactory epithelium: the orthonasal route and the retronasal route. When most people imagine the act of smelling, they picture themselves holding something in front of their nose and inhaling. This describes the orthonasal route. Aroma molecules travel in through your nostrils and up into your nasal cavity to reach the olfactory epithelium. The retronasal route proceeds in a different direction, with aroma molecules traveling from your oral cavity up to your nasal cavity and across the olfactory epithelium. Retronasal olfaction plays a starring role in flavor perception, as it allows you to experience the aroma of a food or beverage while it is in your mouth.

A quick and simple experiment can help demonstrate both the retronasal route itself as well as the impact that aroma has on overall flavor perception. I usually perform this experiment using a jellybean, though you can do it with most types of food. Pinch your nose so as to block any air from moving from your mouth to your nasal cavity. Place the food in your mouth and chew for five to ten seconds, paying attention to the tastes that you perceive. Then, release your hold on your nose and breathe out through your nostrils, noting how the flavor changes. With your nose held, you will primarily perceive taste elements, perhaps noting a sweet taste if you are using a jellybean or some other type of candy. When you release your nose, air travels along the retronasal pathway, carrying aroma molecules from your mouth up to your olfactory epithelium. In an instant, the complete flavor profile comes into focus.

Aside from the different direction the air travels to reach our olfactory epithelium, there is no inherent difference in the way we perceive retronasal and orthonasal aroma. The retronasal route does not engage a different set of receptors or send signals to the brain through a different neural pathway. However, differences in the aromas expressed may emerge when the food or beverage interacts with the environment inside of your mouth. Taking beer into your mouth will cause the sample to warm, subsequently releasing more aroma compounds from the sample. Additionally, while the sample sits in your mouth it interacts with your saliva, which will raise the overall pH of the beer, suppressing some aroma compounds while amplifying others. Finally, we cannot experience aromas retronasally without simultaneously perceiving taste and mouthfeel, which can alter our brain's perception of the aromas (Shepherd 2012, 117). While the orthonasal and retronasal routes target the same neural pathways, the aroma profile of a beer may change between your initial orthonasal assessment and the retronasal assessment that you make once you take the sample into your mouth. In chapter 4, we will cover different aroma assessment methods for both the orthonasal and retronasal routes (p. 43).

THE SENSE OF TOUCH

While most people focus on taste and aroma when discussing flavor perception, our sense of touch plays a significant role as well. Our sense of touch encapsulates a wide range of possible sensations, from pressure and texture to temperature and pain. In the context of food and beverages, we refer to the touch sensations that occur in the mouth collectively as *mouthfeel*. They are also sometimes called trigeminal sensations, because the trigeminal nerve innervates the tongue and much of the face, carrying touch stimuli from the mouth back to the brain for processing. Drinking beer can trigger a wide variety of mouthfeel sensations, each of which impact the overall flavor profile in their own unique way.

Body

One of the key components of mouthfeel is the body of the beer, that is, a measure of the fullness of the beer on the palate. While difficult to define quantitatively, body is influenced by beer density and viscosity. Determining density is a straightforward exercise, only requiring simple measurement of finishing gravity. Higher finishing gravities generally correlate to higher perceived levels of body (Langstaff, Guinard, and Lewis 1991, 431). Beer viscosity is significantly more complex because it depends on a wide variety of different attributes. The levels of protein, glycerol, beta-glucans, polyphenols, dextrins, chloride, and alcohol can all impact the viscosity and the perceived body of the beer (Langstaff and Lewis 1993, 35).

Untrained tasters will often use the term mouthfeel as a stand-in for body, for example, "This stout has a full mouthfeel." While not technically incorrect, mouthfeel describes a number of different tactile sensations experienced in the mouth, while body specifically addresses palate fullness. Encourage your panelists to be as precise as possible when describing and analyzing beers, as this will yield stronger, more useful data.

Carbonation

While carbonation plays a role in taste perception—particularly through the interaction of dissolved carbonic acid with sour taste receptors—it primarily influences flavor perception within the realm of mouthfeel. Technically speaking, carbonation is classified as an irritant based on its mode of interaction with touch receptors in the mouth. However, we typically find this "irritation" pleasant, as evidenced by the vast number of carbonated beverages sought after and consumed by humans. We measure carbonation in beer using either volumes of CO_2 or grams of CO_2 per liter. Volumes turns out to be a very literal measurement. A beer with 2.5 volumes of CO_2 indicates that a given volume of that beer, say, a pint, contains

2.5 pints of carbon dioxide dissolved in it.[5] Beer ranges from approximately 1.0 to 5.0 volumes of CO_2, with most standard ales and lagers falling between 2.3 and 2.8 volumes. Cask beers and nitro beers typically fall between 1.0 and 1.5 volumes, while most German weissbiers and many Belgian styles occupy the higher end of the scale, ranging from 3.0 volumes up to 4.5 volumes or even higher.

These numbers correlate well with the perceived level of carbonation present. A pint of cask bitter will yield barely a tingle on the palate, while an effervescent Belgian golden strong ale (e.g., Duvel) will give a zippy, prickly sensation as it froths across the tongue. Although brewers and consumers alike often focus on the aroma and taste characteristics of a given brand or style, carbonation level plays an integral role as well. Just imagine having a still glass of Duvel or an effervescent cask bitter—the beer would be virtually unrecognizable!

Alcohol Warmth

While carbonation serves as a mild irritant, ethanol is a potent one. If you have taken a shot of high-proof liquor you can attest to the burning sensation produced by the high level of alcohol. At the levels found in beer, you will not encounter this degree of irritation, but higher-alcohol beers can produce warming sensations on the palate and in the throat. Alcohol warmth on the palate is influenced not only by the presence of ethanol but also fusel alcohols. Whereas high levels of fusel alcohols often result in harsh or "hot" notes, low levels of alcohol warmth can offer a pleasant point of complexity within a higher-alcohol beer.

Astringency

We perceive astringency as a drying or puckering sensation on the tongue. In beer, astringency results from elevated levels of tannins or polyphenols. When we drink a beverage high in polyphenols, these compounds bind to our salivary proteins that normally maintain the slippery quality of our saliva. As these protein–tannin complexes precipitate, our saliva loses its ability to lubricate the tongue, producing a drying, rough sensation on the palate.

Several beverage categories feature astringency as a defining trait. Red wines, with high levels of tannins extracted from grape skins, can feature high levels of astringency when young. Tea, deriving polyphenolic tannins from tea leaves, can exhibit significant astringency, with longer steeping times leading to higher levels of astringency. In beer, astringency typically comes from one of two sources: tannins in barley husks or polyphenolic compounds found in vegetative hop material. At low levels, astringent character in beer can lend a complimentary dryness to certain styles of beer. However, at higher levels astringency becomes unpleasant and undesirable, negatively affecting a beer's drinkability.

Metallic

Multiple sensory modalities appear to play a role in the perception of metallic flavor. There is evidence that metal ions can trigger electrochemical touch receptors on the tongue, and that this mouthfeel stimulates and influences the brain's perception of metallic flavor. However, metal ions can also stimulate aroma response, albeit indirectly. Metal ions themselves are not odor active. However, when metal ions interact with fatty acids found on our skin and other tissues they catalyze a reaction that forms highly aromatic compounds that smell strongly of metal. This reaction can occur in our throat after consuming a beer containing metal ions, allowing us to perceive this metallic aroma retronasally (Lawless and Heymann 2010, 46–47).

Certain oxidation characteristics can also mimic metallic aromas (Steinhaus and Schieberle 2000, 1780). We sometimes refer to these compounds as presenting a "false metallic" character, alluding to the fact that these compounds do not result from the presence of metal ions and consequently do not produce any mouthfeel effects.

Capsaicin

Chemical heat derived from capsaicin—the compound responsible for chili pepper heat or spiciness—is one of the better studied mouthfeel sensations. Capsaicin affects thermoreceptors in the mouth, producing a response similar to that of exposure to high heat. This explains why those suffering the effects of spicy food describe the sensation as one of burning, and why cool temperatures can offer immediate relief, if only temporarily.

---

[5]   When measured at standard temperature and pressure (STP), i.e., 0°C and 1 atm.

Capsaicin does not feature in most styles of beer, but as brewers continue to experiment with all manner of novel ingredients the number of chili beers on the market is steadily rising. Depending on the amount used, sensations may range from a mild tickle of heat to an overwhelming burn.

Foam Volume and Quality

The cap of foam adorning the surface of a beer marks it as unique among beverages. Beer foam is a complex topic, and entire books have been dedicated to its structural elements and manipulation of its qualities. At its core, foam consists of a complex colloidal structure made up primarily of malt-derived proteins and hop-derived iso-alpha acids that traps carbon dioxide.

Although some consider foam primarily an aesthetic element, it also influences textural elements of mouthfeel. Further differences arise when considering foam heads atop regularly carbonated beers versus nitrogenated beers. Nitrogenated beers form foam heads with much smaller bubbles, leading to a denser, longer-lasting, creamier cap of foam. As you sip the liquid beer through this dense foam, the head itself lends the mouthfeel a creamy texture. The liquid itself certainly does not elicit descriptions of creaminess. Try spooning the head off of a Guinness and just drinking the liquid—you will immediately notice the immense impact that the foam has on the overall textural experience of the beer.

Other Sensations

Beer can exhibit a wide variety of other textural sensations: the slickness produced by diacetyl and the oily texture contributed by oats are just two examples. These other textural sensations are not well defined within the literature and may be more challenging to pinpoint as specific elements of your beers. However, if one of your beers does feature a specific textural sensation, you may wish to define the sensation within your panel and have panelists evaluate that attribute when tasting that beer.

THE SENSE OF SIGHT

To produce vision, our eyes interpret electromagnetic waves across a relatively narrow band of frequencies, covering wavelengths ranging from about 370 to 730 nanometers. Different wavelengths within this range are perceived as different colors, while the brightness of the light is a function of the intensity of the light source.

Of the senses, humans rely on vision most of all, giving our sense of sight outsized importance when it comes to defining our perceptions. As visual creatures, elements of appearance can drive consumer acceptance of a beer just as much, if not more so, than other sensory attributes. Before we bring a pint to our lips for our first taste, we subconsciously assess its color, clarity, and foam. For much of the past century, a lack of clarity represented a quality issue for most styles of beer, grounds for immediate rejection. Today, a lack of clarity actually drives consumer acceptance of certain styles, as evidenced by the recent rise in popularity of New England IPA. Clearly, aesthetic attributes are not inherently good or bad, but reflect current cultural perceptions and expectations.

Beyond simply influencing consumer preferences, visual cues can also have a very real impact on the perception of flavor, modifying and, in some cases, overshadowing the brain's perceptions of the product's aroma and taste. Most notably, color can shape our perceptions of flavor. Color primarily affects the perceived flavor of a food or beverage in two key ways, either by altering the intensity of a given flavor or by impacting the actual flavors perceived. The scientific literature offers mixed messages regarding the effects on intensity. Some experiments have demonstrated that increased intensity of coloration will increase the perceived intensity of food flavor, but many similarly constructed experiments have failed to reproduce these findings (Spence et al. 2010, 69).

Conversely, the scientific literature shows significant consensus that color can affect the ability to identify a perceived flavor. In experiments featuring liquids flavored with one element and colored in a variety of ways (e.g., liquids flavored with raspberry but colored red, orange, yellow, and green), participants demonstrate significantly higher acuity in correctly identifying the flavor present when paired with the expected color (e.g., raspberry with red color). Furthermore, when the flavor was paired with an unexpected color (e.g., raspberry flavor with green color), participants would sometimes identify the flavor based on appearance rather than on the actual aromas present—in this case finding lime aromas in the green raspberry sample (Spence et al. 2010, 74–75).

A famous study that hits closer to home involved serving white wine dyed red to a panel of tasters. When assessing the dyed wine, panelists used terms common to red wine, such as blackcurrant, raspberry, cherry, prune, and strawberry, and failed to find flavors of banana, pear, and pineapple that they had previously described when presented with an undyed sample of the same wine (Morrot, Brochet, and Dubourdieu 2001, 309). Subconsciously, the panelists ignored actual aroma cues and instead constructed the aroma profile from experience and expectation based on the visual cues presented to them.

While dyeing wine or beer may seem like a dirty trick to play on tasters, these findings have real implications for the way we experience beer flavor. Some beers employ caramel coloring or other artificial darkening agents to elicit aromas of caramel, toast, nuts, chocolate, and other flavors associated with dark malts. At the other end of the spectrum lies the novel concept of a white stout—a blonde beer infused with roasty flavors through the use of coffee or other uncolored ingredients. The result offers a disorienting sensory experience in which the taster attempts to rationalize the cognitive dissonance between what they see and what they smell. Of course, that assumes the taster is in on the ruse. Absent knowledge of the beer's identity, the taster may miss or even ignore the roasty notes present, as the appearance of the beer would otherwise preclude those flavors.

As with taste and aroma, you may encounter variation in the visual acuity of your panelists. While most studies to date have focused on subjects of European descent, the findings suggest that color blindness in particular is a relatively common phenomenon, affecting as much as eight percent of men of northern European ancestry (Wong 2011, 441).[6] Color blindness should not preclude an individual from sitting on a panel, but be sensitive to the fact that it may make evaluation of beer color significantly more difficult for them. Rather than requiring a color-blind panelist to go through the frustrating exercise of trying to assess beer color, allow them to skip any color-related assessments, instead focusing solely on the aroma, taste, and mouthfeel characteristics of the samples they evaluate.

THE SENSE OF SOUND

Of the commonly discussed senses, sound carries the least relevance in beer flavor perception. This is not to say that sound does not have an impact on flavor perception. Various studies have shown that ambient noise can alter the way we perceive certain tastes (Spence 2012, 515). In one particular study, researchers asked subjects to bite down on a number of potato chips while wearing headphones. By adjusting the volume and frequency of the "crunch" heard when a subject bit down on a chip, the researchers were actually able to affect the subjects' tactile impressions of the crunchiness of the chips, with louder or higher-frequency sounds increasing the perception of crunchiness (Zampini and Spence 2005, 347). One can imagine how the "pffft" of gas that accompanies the removal of a bottle cap might build anticipation and lead to heightened enjoyment, but this phenomenon relates to beer service more so than evaluation in the sensory lab. Individual beers offer few auditory cues worth assessing once poured into a glass. Consequently, I will not present you with any specific techniques for assessing the sounds of your beer.

I do, however, recommend that your tasting area be as quiet as possible. You should aim to reduce any ambient noise, as well as the chatter of your panelists while others are tasting, which can both distract and potentially bias your panelists. While low levels of background noise should not significantly affect your results, reducing the overall noise level of your tasting space will allow your panelists to better focus their attention on their other sensory perceptions when evaluating each beer.

---

[6] The most common type of color blindness results from genes found on the X chromosome, resulting in significantly higher rates of color blindness in men. Women of northern European descent experience color blindness at a rate of 0.5%.

3
BIAS AND THE BRAIN

With sensory work, we rely on human subjects to analyze and evaluate our beers. Human sensory data provide us with invaluable insights, particularly when it comes to hedonic (preference) assessment and descriptive profiling with tangible, easy-to-understand descriptive language. However, as humans we struggle with sensory tasks requiring us to report absolute measurements of a given quality or characteristic—we are far more adept at performing comparative judgments (Lawless and Heymann 2010, 204). Given two different objects, you could probably correctly identify which one weighs more, but you turn to a scale if you need to know the exact weight. When we assess the level of bitterness in a beer, we compare the beer in front of us against our memories of other beers, memories which are prone to error. And because sensory evaluation requires a number of different processing steps within our brains, we open ourselves to the effects of a wide variety of biases that can distort our final perception of flavor.

Biases often creep into our cognitive processes without our knowledge, silently and subconsciously influencing the way we interpret our own personal realities. From the way we digest current events to the way we assess event probabilities, cognitive biases subtly shape our perceptions each and every day. Interestingly, we are often quite adept at spotting biases that affect other people's judgment while being almost comically unaware of our own (Scopelliti et al. 2015, 2468). Even if we manage to recognize one of our biases, either through introspection or someone else bringing it to our attention, our brains often cannot entirely shake its effects.

While not exactly biases themselves, optical illusions mimic the way that cognitive biases affect our perceptions. Take the classic Müller-Lyer illusion—even after you have seen the trick and know that the lines are the same length, one still appears longer than the other. You can measure the lines or cover up the fins of the arrows to establish with one hundred percent certainty that the lines are identical, but when you step back and look at the image as a whole your brain still falls for the trick.

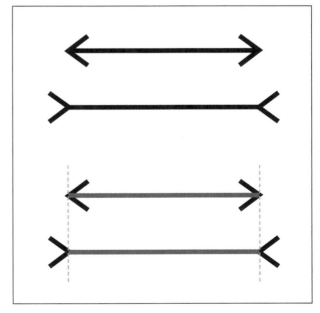

Figure 3.1. The Müller-Lyer optical illusion.

While recognizing the possibility of a particular bias will not entirely eliminate its effects, it does allow us to limit the impact that it has on our perceptions. In some cases, we can develop techniques to help panelists circumvent certain biases. In other cases, careful design of tasting sessions will help eliminate potential sources of bias for your panelists. Though you can never fully remove the impacts of bias, teaching your panelists to recognize biases in their own tasting can help them enormously, leading to higher acuity and better results. Through your efforts to limit sources of bias, you will ultimately reap the rewards in the form of better, more accurate data.

HOW BIAS OCCURS

In every waking moment, our brains process a constant, endless stream of different sensory stimuli. Our senses perpetually monitor the world around us—the ambient noises of our environment, the temperature, what the person across the room is doing out of the corner of our eye. If we were to somehow try to manually process all of this incoming information at all times, it would leave us paralyzed, all of our processing power dedicated to simply taking in information with nothing remaining to actually make decisions or take action. Fortunately, the brain excels at processing sensory information, filtering out static signals such as the ambient temperature, the pressure sensations of your body against a chair, or the background noise of your surroundings, instead primarily concerning itself with registering changes in your sensory environment (Lawless and Heymann 2010, 206). We also possess the ability to focus on specific sensory inputs out of a sea of possibilities. The so-called "cocktail party effect" describes our capacity to focus on our exchange with the person in front of us while ignoring the multitude of conversations occurring all around us (Johansen-Berg and Lloyd 2000, 894).

Bias can arise for a number of reasons. One common explanation stems from the brain's use of heuristics, that is, the use of mental shortcuts or rules of thumb that simplify decision-making (Kahneman 2011, loc. 84 of 9418). The brain relies on heuristics to cope with the inherent limitations in processing the sheer volume of incoming stimuli, but this reliance sometimes results in errors in judgment or thinking, which are known as cognitive biases (Buss 2005, 725). While rooted in the field of psychology, several types of cognitive bias have made their way into common vernacular due to their prevalence. Some biases are rather frivolous, more humorous than consequential.

The planning fallacy, for example, describes the cognitive bias in which people regularly underestimate the amount of time that a given task will take (Buehler, Griffin, and Ross 1994, 366). While falling victim to the planning fallacy may frustrate those around you, its ramifications typically end there.

However, biases can have large impacts on our lives, in some cases shaping the way we view the world. Confirmation bias, in which we seek out information that supports or validates our beliefs, allows us to live without constantly questioning our opinions. While this can serve as a boon in some scenarios, confirmation bias shows its negative side when we are wrong about a given fact or belief, often preventing us from correcting the error even when presented with accurate information. Confirmation bias often undergirds the deeply entrenched positions of different parties arguing over contentious political or social issues and usually leads both groups to harden their stances following a disagreement.

When processing flavor, our brains attempt to weave together a multitude of different sensory inputs into a single cohesive output. A lifetime of subconsciously processing flavor inputs causes us to build up a number of shortcuts and biases, which can have an impact on panelists' decisions and the data they produce. Whether discussing broader biases that affect our views of the world or simpler biases that primarily come into play during a sensory session, noting these biases can help to diminish their effect. Our brains will still attempt to take subconscious shortcuts but, if we pay attention, we can sometimes catch our brains in the act, allowing us to adjust for biased judgments as they occur.

CONFIRMATION BIASES

Confirmation biases form one of the largest and most pervasive groups of biases. In popular usage, confirmation bias often comes up within the context of large-scale systemic issues, such as political beliefs or racial prejudices. However, confirmation biases can also influence less consequential scenarios, such as an individual flavor perception. When tasting a sample, if we hold a certain belief about that sample, our perceptions and our interpretation of those perceptions will often follow. Confirmation biases represent some of the more insidious biases, in that they are very difficult for individuals to ignore even once alerted to their existence.

Expectation Error

Expectation error describes a taster using previous knowledge of a sample to form a sensory judgment, rather than using their actual sensory perceptions (Meilgaard, Civille, and Carr 2016, 46). If some of your panelists work in production, their knowledge of what is going on in the brewhouse can lead to biased judgment. For example, if a panelist knows that the most recent batch of IPA experienced a sluggish fermentation, they may note fermentation flaws in a sample of the IPA, even if it does not show those flaws and even if it happens to be a sample from a different batch of IPA! They will search for sensory cues and may even encounter supporting perceptions to back up their expectation, even if those attributes do not actually appear in the beer. Tasting shelf-life samples of beer also presents difficulties related to expectation error. Because panelists anticipate the beer will exhibit aged characteristics, they may note high levels of oxidation flavors regardless of the actual profile of the beer.

Selective Perception

A corollary to confirmation bias, selective perception describes one way in which our brains deal with information that does not fit with our prior beliefs or knowledge. Through selective perception, we either fail to notice, quickly forget, or outright ignore information that falls outside of our expectations or beliefs (Griffin 2013, 259). A panelist might overlook a flavor cue that they were not expecting, for example, missing a flaw on a true-to-target test for a beer that they have tasted numerous times before. Alternatively, a panelist performing a descriptive test might assign less importance to a characteristic that was not supposed to be there, ultimately omitting it from their assessment of the beer. A panelist tasting a shelf-life sample might focus solely on aged flavors, causing them to miss notes of hop character that would otherwise indicate that the sample was still relatively fresh. The panelist's ignorance of these unexpected traits is neither deliberate nor nefarious—oftentimes this process occurs subconsciously, as the brain gives greater credence to our expectations than our actual sensory experience.

Avoiding Confirmation Biases

To limit expectation error and selective perception, avoid giving panelists any unnecessary information about the sample in question. When presenting samples for hedonic or descriptive ratings, make sure that panelists are entirely blind to the identity of the sample. Although you might think that a well-trained panelist would not allow information about a sample to bias their judgments, in reality their brain will exploit any details available and use them to reach a conclusion.

You should occasionally incorporate spiked or doctored samples into your panel sessions, or serve panelists the incorrect beer on a product release panel, for example, swapping your IPA out for a similar brand, either from your lineup or from another brewery. Do not approach this with the goal of trying to trick your panelists or catch them making a mistake. Rather, the knowledge that such a spiked or "trick" sample may appear on any given panel will keep panelists on their toes and force them to make decisions based on their actual sensory perceptions, helping them to avoid confirmation biases and other biases.

Ultimately, panelists are human, and as humans we have a strong desire to be right when presented with a test. Your panelists will derive pride from their ability to correctly assess samples during panel sessions. As a result, their brains will use any information—relevant or irrelevant—presented to them in an attempt to arrive at the correct answer. You should always strive to present samples in a uniform fashion and to limit any cues that could influence panelist decisions. Present samples without any identifying information beyond what is explicitly required for the test. Similarly, avoid calling emergency panel sessions if possible, as this will prompt panelists to expect flaws in the samples tasted. Primed with the knowledge that something might be amiss, panelists will be more likely to report defects even if none are present.

Mutual Suggestion

You are sitting around, enjoying a new IPA with friends, puzzling over its beguiling aroma, when one of your friends interjects, "I love the passion fruit notes in this beer!" Like the sun emerging from behind a cloud, an image of passion fruit bursts into your mind, clear as day. "That's it!" you think, the flavor plainly obvious to you now. . . Or is it? Perhaps the passion fruit flavor was there all along and you simply failed to recognize it at first. However, it is equally likely that your brain conjured the perception of that flavor upon hearing it spoken aloud, altering your flavor experience.

Mutual suggestion occurs when one taster's opinion of a beer influences the opinion of other tasters (Meilgaard, Civille, and Carr 2016, 49). Although not exactly a confirmation bias, mutual suggestion can trigger an expectation error, causing us to anticipate the presence of the named flavor and subsequently find evidence of that flavor, whether or not it actually exists. And once you have heard a flavor note suggested by another panelist, you cannot ignore it or prevent it from shaping your perception. Trying to exclude it from your evaluation can cause just as many issues as accepting its presence.

Humans are social creatures. When assessing beer, we naturally want to share our experience with our fellow tasters, but to avoid biasing other panelists we must resist this urge. To limit the potential for mutual suggestion during a session, panelists should not talk with one another during their initial tasting. Even making a face or a noise of disgust can bias the results of other panelists. In traditional sensory settings, panelists evaluate samples in individual booths, which prevents them from interacting with the other panelists. However, if you do not have the luxury of a dedicated sensory space with booths, you can arrange panelists at tables along the wall to prevent them from interacting with one another. Sitting panelists around a table risks the potential for mutual suggestion. Even the most fastidious panelist will likely find it difficult to resist the temptation to exchange a knowing glance with a fellow taster upon encountering a spiked sample. Using a single table with panelists seated in a circle will still produce useful data, but be aware of the potential pitfalls. If possible, you should first attempt other configurations.

INFORMATION ERRORS

Information errors form a group of biases covering flaws in judgment that occur when a panelist subconsciously uses a piece of information or knowledge to impact decisions, rather than relying on their perceptions. This includes errors in which determinations within one sensory modality or beer characteristic affect assessments of other, unrelated characteristics.

Logical Error

A logical error occurs when a taster builds an association in their mind between two distinct characteristics (Meilgaard, Civille, and Carr 2016, 47). Two common examples include fruity aroma matched with sweet taste, and dark color paired with roasty aromas. But beer exhibits a vast number of other potentially associated traits. For example, tasters often form a subconscious association between the level of hop aroma and the level of hop bitterness found in a beer. Tasters have been shown to report higher levels of perceived bitterness as the hop aroma of a beer increases, even if the measured level of bitterness does not actually change (Oladokun et al. 2016, 104). Each of these logical errors involves cross-modal sensory misfires, resulting from tasters conflating two distinct characteristics that typically appear together.

You can help panelists combat logical errors through a combination of awareness and training. Upon recognizing common logical error traps that we fall into it becomes easier to spot them in the future and limits their influence. If we know that fruity aroma does not necessarily correspond to increased sweetness, we can take extra care to evaluate those two attributes separately. Furthermore, we can use mechanical techniques, such as closing our eyes to block sight or pinching our noses to block aroma, to help limit cross-modal interactions.

If you find that logical errors present a significant issue for your panelists, training on individual attributes may help. Presenting panelists with beers featuring fruity aromas and different levels of sweetness (you might doctor the sweetness level yourself) can help panelists separate the two attributes from one another. Coloring a pale beer with dark Sinamar® dye will teach panelists to evaluate appearance separately from aroma and taste.

Halo Effect

The halo effect describes an error in which positive evaluation of one characteristic boosts the rating of other independent characteristics. Many classic psychology studies of the halo effect focus on how individual physical or personality traits can impact our global perceptions of other people, for example, subconsciously elevating our estimation of a person's intelligence and friendliness if we find them attractive (Nisbett and Wilson 1977, 250). However, the halo effect can play a significant role in sensory judgments as well. In an experiment demonstrating the halo effect, researchers presented subjects with low-fat milk supplemented with a near-threshold amount of vanilla extract. At this level, vanilla extract will alter the perceived aromas of the milk but will leave both taste and mouthfeel unchanged. However, this study found that the addition of vanilla extract not only increased the overall liking scores for the low-fat milk, but also increased subjects' ratings of sweetness, thickness, and creaminess. Although none of these attributes actually changed, tasters' liking of the product went up, leading them to rate other positive attributes more highly (Lawless and Heymann 2010, 216).

Halo effects can arise during sensory evaluation, beer judging, or even just casual enjoyment of beer. Given our visual nature, various appearance attributes can produce potent halo effects. The brilliant clarity of a well-made German Pilsner might boost a taster's rating of its crisp, dry finish or its snappy bitterness. Conversely, the hazy sheen of a New England IPA plays up our expectations of hop character, increasing perception of tropical fruit and citrus notes, so increasing overall enjoyment of the product.

While some sources use the term halo effect to describe both positive and negative reactions, some texts limit the halo effect to positive contexts. The opposite effect will sometimes be called a "horn effect," occurring when a negative rating of a given attribute brings down the score of other unrelated parameters. In the case of a New England IPA with substandard haze, a consumer may rate the aroma as lackluster, regardless of how much hop aroma the beer presents. If a panelist encounters an off-aroma in a beer, it may cause them to rate other aspects of the beer poorly as well. Notes of acetaldehyde on the nose might lead a panelist to find issues with the taste and mouthfeel, even if those characteristics present without flaws.

You can help panelists limit halo and horn effects by explicitly teaching them to evaluate distinct attributes separately from each other. Outside of hedonic assessments, panelists should split evaluation of a sample across the different sensory modalities of appearance, aroma, taste, and mouthfeel. Each of those modalities can be further subdivided. For example, panelists should assess sweetness, bitterness, and acidity individually when tasting a sample. In dissecting mouthfeel, encourage them to examine body, carbonation level, alcohol warmth, and astringency one at a time rather than all at once. By focusing on each attribute independently your panelists will greatly improve their accuracy and simultaneously limit potential halo effects.

PRESENTATION ORDER BIASES AND EFFECTS

Sample order can greatly impact our perceptions, whether we are arranging samples for presentation to panelists in the sensory lab, ordering the beers within a flight in a competition setting, or simply switching between beers while drinking in the taproom. In most cases, careful planning and proper instruction can help to mitigate these issues, but you should always keep sample order effects in mind when planning a tasting.

Contrast Effect

The contrast effect occurs when two samples presented sequentially differ in one or more parameters. Our brains, adept at recognizing changes, will tend to overestimate the magnitude of the difference (Lawless and Heymann 2010, 206–9). As a result, if you present a high-bitterness beer immediately prior to a low-bitterness beer, tasters will rate the bitterness level of the low-bitterness beer as less bitter than they would have if they had tasted that beer on its own. This is not a physical acclimation or desensitization effect, but rather a psychological consequence of the way our brain processes contrasting stimuli. In the reverse scenario where the low-bitterness beer is presented first, the high-bitterness beer will actually taste more bitter than if it had been tasted in isolation.

This effect exhibits greater influence on hedonic tests that measure the general liking of a beer. Presentation of a "poor" sample prior to a "good" sample will tend to result in a higher rating for the good sample, whereas presentation of a good sample prior to a poor sample will result in a lower rating for the poor sample (Meilgaard, Civille, and Carr 2016, 48). This can have important implications within the context of judging beer, inherently a somewhat hedonic exercise. If each judge tastes the beers in a given flight in the same order, some samples may receive artificially inflated scores simply because they were preceded by an unpleasant sample. At large competitions like the Great American Beer Festival® and the World Beer Cup®, stewards serve judges all of their samples at once, so the competition organizers encourage judges to randomize the order in which they taste the beers to prevent contrast effects from affecting their verdicts. Similarly, you can employ randomization of sample order to help improve the quality of hedonic data from your panel.

ADAPTATION EFFECTS

Adaptation effects do not fall within the same realm as cognitive biases since they often stem from features of the sensory systems themselves, rather than resulting from the way the brain interprets sensory input. Nonetheless, adaptation effects can similarly distort panel data and lead to flawed decision-making, so they must be carefully considered when planning your sample presentation order. Sensory scientist Lindsay Barr shared an anecdote from her time leading the sensory program at New Belgium Brewing, in which a large number of panelists failed a batch of Ranger IPA because of low hop aroma. Smelling the sample herself, Lindsay found the hop aroma in line with her expectation for the brand, so she took a closer look at the setup of that day's session. She found that, in this case, the Ranger IPA had been served immediately following a sample of another hop-forward brand, Citradelic Tangerine IPA. Lindsay surmised that adaptation to the hop aroma of Citradelic had led to a skewed assessment of the Ranger IPA. She tested this theory at the next panel session by running the same batch of Ranger IPA placed in a different position within the flight. The sample passed easily, with a majority of panelists marking the sample true-to-target.

Even experienced panel leaders will occasionally misjudge the way their beers will behave when served in sequence. While she had previously thought of Citradelic as less intense than Ranger IPA, Lindsay found that she could not serve the two back-to-back, and updated sample order protocols to reflect this finding. Critical examination of the test structure and presentation order of samples prevented Lindsay from sounding the alarm on a batch of perfectly good Ranger IPA.[1]

Pattern Effect

Pattern effects can occur when a panelist detects a general pattern to the order of the presentation of samples. Upon detecting a pattern, the panelist's brain will exploit that information to create an expectation error, in which the panelist's perceptions match the perceived pattern. In most cases, we try to present samples in order of increasing intensity. When administering a true-to-target panel in which panelists know the identities of the samples they are tasting, this does not pose a problem. However, if you always present samples in this manner when administering hedonic or descriptive panels, this sort of pattern can sway panelists' verdicts, resulting in them reporting earlier samples at lower levels of flavor intensity and later samples at higher levels (Meilgaard, Civille, and Carr 2016, 47).

First Sample Effect

Studies have repeatedly shown that tasters tend to assess the first sample of a flight more favorably than the rest of the samples presented within a group (Dean 1980, 107), though the effect appears to be diminished in experienced tasters (Mantonakis et al. 2009, 1309).

Discrimination between samples in a difference test is also highest with the first set of samples presented within a session, though this likely reflects the fatigue and diminished focus that panelists experience in later sample sets (Meilgaard, Civille, and Carr 2016, 48).

Limiting Presentation Order Effects

Most fixes designed to limit the impact of presentation order biases do not reduce bias within the individual, but rather attempt to mitigate the bias across a group of tasters. The most common solutions use balanced or randomized serving orders. While randomized serving order presents the samples to each panelist in a truly random fashion, balanced presentation presents every possible order an equal number of times. For example, in a flight with three samples, six different presentation orders are possible (fig. 3.2). If presenting these samples in a balanced manner to a group of 24 panelists, each order would be seen by four panelists. Samples presented in a balanced manner still leave room for individual panelists to experience contrast effects or first sample effects, but since other panelists receive the samples in a different

---

[1]   L. Barr, telephone conversation with author, February 25, 2020.

A	B	C
A	C	B
B	A	C
B	C	A
C	A	B
C	B	A

Figure 3.2 – Three distinct samples, A, B, and C, may be presented to panelists in six possible arrangements.

order, the aggregate data of the group should be free of any presentation order effects.

However, depending on the size of your group or the structure of your panel, it might not be feasible to serve samples this way. If you cannot present samples in a randomized or balanced manner, you can still take steps to mitigate the impact of these effects. You can diminish first sample effects to some extent by serving panelists a warm-up beer to taste before they begin the session, or even by beginning the panel session with evaluation of a dummy sample that does not factor into your data set. You can limit pattern effects by consciously designing hedonic or descriptive panels to avoid a predictable order or by using a random number generator to determine the presentation order each time you set up a panel. Varying presentation order will also help reduce panelists' complacency, lessening the danger of default responses.

OTHER BIASES AND EFFECTS

Default Effect

When given a choice between two or more options, panelists will usually favor the default response if one exists, particularly if they become fatigued (Danziger, Levav, and Avnaim-Pesso 2011, 6889). This default effect presents difficulties in true-to-target testing. If panelists find themselves always passing samples through, they may become complacent in their evaluation, causing them to miss flawed or aberrant samples. Like several other biases and effects, you can combat default effects by occasionally spiking true-to-target samples to ensure that your panelists will, in fact, fail samples that fall outside of the brand's specifications. You do not need to spike beers on every panel; depending on the frequency with which

you hold panel sessions, I recommend including spiked samples somewhere between once a week (if conducting panels on a daily basis) and once a month (if conducting panels on a weekly basis). For spiked samples to be effective at keeping panelists focused, you should employ them randomly, without pattern. If panelists detect a pattern (e.g., a spiked sample shows up in the panel session every other Wednesday), they will slip back into default mode, opening them up to a variety of potential biases.

IKEA Effect

The IKEA effect describes a bias in which people tend to place higher value on things that they have created (Norton, Mochon, and Ariely 2011, 453). In the context of beermaking, this can mean that brewers do not always make the best panelists, in part because they may subconsciously overlook faults or inflate their rating of a beer due to their involvement in its production. This is not to say that brewers should not participate on sensory panels—on the contrary, brewers can be exceptional panelists due to their depth of production knowledge. However, you should try to counterbalance their participation by including employees from outside of the brewing team.

Lack of Motivation

Though not a bias, panelist motivation levels will greatly affect their performance, as panelists lacking motivation will not devote the entirety of their attention to the task at hand. Consequently, their acuity suffers and the validity of their results diminishes. Every brewery approaches panelist motivation in a slightly different way, but I have compiled some of the most effective methods for maintaining panelist motivation in chapter 9 (p. 116).

PANEL LEADER BIASES

The biases presented in this chapter primarily describe sensory errors that can occur when panelists taste beer. However, panel leaders are not immune from bias. Many of the same biases that afflict panelists can have a negative impact on a panel leader's judgment, albeit in different ways. As just one example, halo effects can cause you to treat data from different panelists differently. You might assign greater significance to the evaluations of the company president than one of the taproom employees, even though their relative positions within the company have no bearing on their sensory abilities.

Panel leaders must think like scientists in designing their experiments, open to whatever results their panel produces and accepting of the fact that they may be wrong from time to time. Lindsay Barr's anecdote from New Belgium illustrates this concept perfectly (see "Adaptation Effects" sidebar). In the face of her panel delivering a result that did not comport with her own sensory experience, confirmation bias may have caused Lindsay to disbelieve her panelists. Had she succumbed to the IKEA effect, she may have struggled to look critically at the way she had designed that morning's panel. Instead, Lindsay adopted an attitude of curiosity. Confronted with an unexpected outcome, she took a step back and thought critically about what might have occurred, which allowed her to identify the true cause of the issue.

I wanted to close by highlighting one final bias: the information bias. When attempting to make a decision, humans will often seek out as much information as possible, even if that information will not actually inform the decision (Vaughn 2013, 25). The crux of information bias is the belief that more information leads to better decision-making, but even casual examination of this belief causes it to fall apart. Information irrelevant to the matter at hand obviously does not help us make better decisions—it makes it harder to pull out the data that actually matters while wasting time and resources collecting superfluous data. When structuring panel exercises, avoid seeking information that does not have the potential to inform decisions. If you always have an eye towards how you will use the data generated by each test you run, your sensory program will be well positioned to serve the needs and goals of the brewery.

4
SENSORY EVALUATION TECHNIQUES

Although we all use our senses every day to experience flavor, few untrained tasters have taken the time to learn how to harness their senses to evaluate flavor. Regardless of the exact structure of your panelist training, you should begin by teaching proper sensory evaluation techniques, as this will provide panelists with a strong foundation from the get-go. Starting with a robust toolkit of different evaluation techniques will boost panelists' confidence and improve both their acuity and their consistency from session to session.

Lab technicians require training on the proper techniques used with each instrument to safeguard against collecting flawed data. Whether measuring the gravity of wort, taking a pH reading of a fermentor sample, or counting yeast cells from a slurry, incorrect technique will yield inaccurate results. Each procedure follows a specific set of protocols to guarantee accurate outcomes, regardless of who is operating the equipment.

The same principles hold true within the sensory lab. By teaching your panelists how to properly use their senses, you will also reduce the variability between the responses of different panelists. Humans palates are inherently variable; due to genetic and behavioral variation between panelists, some differences in perception should be expected, particularly within the realm of aroma. However, making sure that your panelists evaluate beer using the same set of techniques will help align their responses to some extent, ultimately yielding better data.

TYPES OF SENSORY DATA

Once you have your panel up and running, you should focus on quality control by using true-to-target tests to monitor packaged beer for release (see p. 73). The specifics of the test will be covered in depth in a future chapter, but the method has panelists compare a sample against a brand target and asks a binary question: is the sample true-to-target (TTT) or not true-to-target (not TTT)? This deceptively simple test will provide you with a powerful tool to monitor the consistency of your outgoing beer. But before you can perform true-to-target tests with your panel, you will first need to define the targets for your brands. You will work with your panelists to develop these targets using a description test (see p. 87) to capture the key characteristics of each beer and synthesize your panelists' responses into succinct, holistic brand profiles.

When describing the different characteristics of a beer's profile, your panelists will report their responses using a few different data formats. I shall describe the common formats below using their technical terms but do not be put off by the jargon—you already likely use these different scale types to classify data in both beer evaluation and your everyday life.

The first data type, nominal data, assigns named labels rather than quantitative values to a given parameter. Color provides a straightforward example of a nominal scale. If I presented you with a banana, a strawberry, and a lemon and asked you to assess the color of each, you would report back yellow, red, and yellow, respectively. Nominal data groups items into categories; for example, we could state that both bananas and lemons are yellow fruits. Due to their nature, nominal data do not offer any quantitative information, and you cannot set up any sort of comparative relationship based on nominal data. For example, it does not make any sense to state that red fruits have more (or less) color than yellow fruits. In beer, one of the few attributes reported in a nominal format is beer color, which can be described using simple color words.

Nominal data also applies to parameters assessed using a check-all-that-apply (CATA) format of collection (fig. 4.1). When assessing aroma, your panelists will use a modified CATA format, allowing them to select various aroma descriptors from a previously agreed upon lexicon. Due to the complexity of aroma and the enormous range of possible aromas in beer, it does not make sense to have panelists utilize a giant checklist comprising a hundred or more descriptors. Instead, panelists should assess aroma by providing a short list of the most prominent aromas they find. By aggregating your panelists' responses, you can derive useful insights into the aroma profile of a beer.

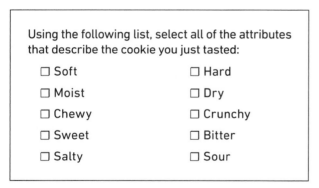

Using the following list, select all of the attributes that describe the cookie you just tasted:

☐ Soft ☐ Hard
☐ Moist ☐ Dry
☐ Chewy ☐ Crunchy
☐ Sweet ☐ Bitter
☐ Salty ☐ Sour

Figure 4.1. An example of a survey question using a check-all-that-apply (CATA) format for data collection.

Sometimes, simply noting the presence or absence of a given characteristic will not provide enough information on that parameter. For example, if describing the taste of different beers, stating that a given sample presented sweetness and bitterness does not tell you much—all beers exhibit some amount of sweetness and bitterness. In these cases, we turn to ordinal data. Ordinal data organizes samples into groups based on different levels of a given characteristic. In other words, an ordinal scale allows us to measure whether bitterness is "low," "medium," or "high." When performing a description test, your panelists will use ordinal scales to measure most beer attributes.

An ordinal scale can be segmented into any number of groups, though convention usually suggests an odd number of categories. When deciding how fine of a scale to use, you need to weigh two competing considerations against one another. On one hand, how granular of a difference do you want to measure in this attribute? Is it sufficient to know that the sweetness level is either low or medium, or do panelists need to report whether it is medium-low or medium? You should also consider that increasing the number of groups on a scale will make it more challenging for panelists to repeatably produce their results. If you use a seven-point scale for sweetness, can your panelists accurately and consistently label a sample as very low, low, or moderately low? In most cases, three (low, medium, and high) or five (low, medium-low, medium, medium-high, and high) categories will suffice. For certain attributes, such as astringency or alcohol warmth, you may also wish to add a "none" category.

True descriptive analysis uses a data format called interval data, a more rigid, quantitative form of ordinal data obtained by grading a specific attribute along a continuum with each scale point separated by equal intervals. For example, if measuring sweetness on a five-point intervallic scale, an increase in perceived sweetness from a 2 to a 3 would indicate the same degree of increase as a 3 to a 4. Intervallic data allows for detailed quantitative analysis, but using the scales precisely requires hundreds of hours of panelist training. While useful in some sensory settings, your brand targets will not require this level of precision; in fact, defining targets too precisely can have negative consequences. Sticking with nominal and ordinal data formats for descriptive testing will allow you to build robust, usable brand targets without requiring an astronomical amount of panelist training.

ASSESSMENT TECHNIQUES

While each sensory technique presented in this chapter is necessarily unique to the sensory system it serves, many of the methods share a common goal: to isolate perception of that sense as much as possible. In the sensory booth, we endeavor to record our sensory observations with a high degree of accuracy, which requires that we break down our experience of a beer's flavor into its component parts. In doing so, we fight against the natural tendency of the brain to present flavor as a single perception, instead carefully picking apart the layers to observe each element of the beer one at a time. In many cases, we achieve this by mechanically blocking off one or more of our other senses, for example, closing our eyes to limit the impact of sight or pinching our noses to limit the influence of aroma. Although panelists may initially balk at these techniques, you should remind them that their goal in sensory evaluation is not to enjoy the samples but to serve as a measuring instrument.

Successfully convincing panelists to use these techniques has a lot to do with how you present these methods. First and foremost, if you introduce these techniques as best practices and standard protocol rather than optional ways to interact with a sample, your panelists will likely follow your lead and buy in to the techniques. Holding your nose and closing your eyes while tasting a beer may feel silly at first, but if you emphasize from the beginning the importance of generating accurate observations and explain that applying these techniques will help the panel achieve valid data, that too can increase adoption.

In addition to isolating each sensory modality, you want your panelists to break apart their assessments of individual attributes within each sensory modality; again, this improves the accuracy of their responses. When assessing taste, instruct them to focus on assessing sweetness, then bitterness, then acidity, rather than tasting the sample once and trying to make a judgment on multiple parameters. By asking yourself the question, "What is the bitterness level of this sample?" you are far more likely to produce an accurate, reproducible result. While this may seem tedious, with practice your panelists will learn to move through the assessment of each beer in a rhythmic, routine fashion, making each assessment within a few seconds and only spending one to two minutes total on each beer. By giving your panelists a framework of simple questions to work within, you simplify the process of evaluating a beer into a series of discreet tasks, which should make things easier for your panelists.

The following sections detail a number of different parameters within each sensory modality that you may wish to have your panelists assess when performing a description test. The specific attributes you choose to track are ultimately up to you but they should reflect the most important characteristics of your beers, particularly ones that you think might impact consumer acceptance or rejection if they vary too far from the standards set for each brand.

TESTING PANELISTS

When training panelists on the attributes of various modalities, whether aroma, taste, or even appearance, you should test them on their ability to assess each attribute. While these tests will help assure you that your panelists can accurately recognize and identify a specific aroma or taste, testing offers a deeper intrinsic value. Countless studies have shown that the human brain most strongly encodes information when forced to retrieve that information. In other words, testing someone on information that they have learned causes them to retain that information better than any other teaching format (Brown, Roediger, and McDaniel 2014, 29).

Many people when studying for a test will reread textbooks or notes in order to prepare. However, this type of preparation often causes us to overestimate mastery of the subject matter. We grow familiar with the way the text presents the information and our brains begin to disengage, comfortable in the knowledge that they do not need to work to interpret the incoming information. Consequently, we do not retain information processed this way very well. In essence, we are practicing reading when we should be practicing recall.

The same principles apply when presenting panelists with training samples. While you should begin training for a specific attribute or characteristic by giving panelists samples or standards to learn from, you should perform blind tests with those standards from the first exposure. This sort of blind testing more closely mimics the exercises that panelists will perform when they sit to assess samples during a sensory session. And beyond the practical application of this format, forcing panelists to perform recall will actually improve their knowledge retention, helping them identify specific flavors or other attributes in future samples with better acuity and greater confidence.

Visual Assessment

Assessment of a sample of beer typically begins with examination of the beer's appearance. Some schools of thought argue that aroma must be assessed immediately upon serving a beer, asserting that some of the more volatile and delicate aromas will dissipate shortly after the beer is poured. Notably, this contention seems to feature in some judging circles driven by score-sheets that place aroma assessment above appearance assessment. However, most aroma compounds remain stable within a beer over the course of a standard evaluation period. In truth, aromas that appear to dissipate quickly often result from adaptation of the taster rather than the compound actually leaving the beer. As such, it makes the most sense to begin assessment with appearance, as we naturally interact with a beer by first noting its visual elements before engaging any of our other senses.

Unlike the other senses, we do not have many specific techniques at our disposal when it comes to assessing the appearance of a sample. However, when creating an environment for sensory evaluation, try to provide your panelists with a well-lit space, ideally with access to a white background to aid in color assessment. Aim for something similar to the level of light found in a standard office space and you should be fine (Meilgaard, Civille, and Carr 2016, 35).

Appearance can be broken down into several distinct attributes. When assessing appearance, your panelists should always evaluate color and clarity, regardless of the beer. The exact parameters you track beyond that will largely depend on the types and styles of beers that you produce. For example, some beers feature a large, well-retained head as a defining characteristic of their appearance. In such cases, you could include head size and retention in the description of your brand target.

Color

While we can quantify color with an SRM or EBC number using instrumental methods, panelists should report color in a nominal format, using simple color adjectives in their descriptions. Work off of a shared, straightforward lexicon. For standard beers, use the terms straw, yellow, orange, amber, brown, and black, allowing panelists to modify each term with light or dark if desired (e.g., light yellow or dark amber). For

beers that fall outside of these terms, stick to standard colors; for example, red, pink, and purple may come into play when using certain fruits. While certainly not the most poetic or evocative language possible, using well-defined color terms will help your panelists achieve consensus, preventing you from having to parse whether burnt sienna, mahogany, and garnet represent distinct colors or just different panelists' ways of describing the same thing.

Clarity/Turbidity

Panelists should measure clarity using an ordinal scale.[1] While you can train your panels on discrete levels of haziness by doctoring samples with specific amounts of tannic acid, you can achieve the same effect by simply presenting panelists with commercial examples exhibiting different amounts of turbidity. However, if one of your core brands features haziness as a key characteristic (e.g., a New England IPA or a witbier), you may want to use a more quantitative scale. Purchasing a turbidity meter represents a significant—and probably unnecessary—investment for most breweries. However, applying a bit of creativity can yield similar results at a fraction of the price.

No stranger to hazy beer, Allagash Brewing Company has been brewing their flagship witbier since day one. In the early days of implementing a quality program, brewmaster Jason Perkins would hold unlabeled bottles in front of a letter "A" (naturally) displayed on his computer screen and gradually increase the font size until he could see the letter through the beer; he would record the final font size as a quantitative measurement of the beer's level of haze.[2] These days, Allagash has upgraded to a more formal method, using a benchtop turbidity meter to measure their beers. However, Jason noted that the bootstrapped computer screen method gave them reliable, useful data. I love this story, as I feel it accurately encapsulates both the ingenuity and scrappiness of modern craft brewers. Simultaneously primitive and elegant, this novel method can offer you quantitative measurement of turbidity at virtually no cost.

Particulate

Haze refers to a uniform turbidity, whereas particulate describes visible solids (sometimes colloquially called "floaties"). Particulate matter can result from all

---

[1]   Remember, any characteristic calling for an ordinal scale just means you should use terms like *none, low, medium,* and *high.*

[2]   J. Perkins, telephone conversation with author, May 7, 2020.

manner of solids in the sample, including yeast, hop material, trub, or other sources. If you choose to track particulate levels, panelists should measure levels using an ordinal scale.

Head Color

For head color, as with beer color, a limited set of simple color descriptors should suffice. Choose words that reflect the head colors across your beers. A common set might include white, light brown, brown, and dark brown. As with all scales, avoid overwrought descriptions and unnecessary specificity. Asking panelists to differentiate between eggshell and ivory goes past the degree of detail needed to maintain your brands, not to mention it also probably extends beyond the discrimination ability of your panelists.

Head Size and Retention

Asking panelists to assess head size and retention requires that you implement strict sample preparation protocols. If you gently pour a beer into a glass with extreme care you can greatly reduce the size of its head, even with a highly carbonated brand. However, pour that same beer straight down the center of the sample glass and you will produce a large, voluminous head. You should determine a set technique for pouring samples and then endeavor to pour each beer in exactly the same manner. Similarly, if samples sit for some amount of time before a panelist receives them then the foam will collapse and carbonation will begin to dissipate. While the logistical realities of running a sensory panel dictate that you have to make some compromises, you should make every effort to consistently present panelists with samples in an identical fashion from sample to sample, day after day. (For more information on sample preparation, see Sample Preparation, p. 102.)

A panelist's technique will also affect their rating of head size and retention. If a panelist swirls the glass before assessing these parameters, their swirling technique will impact the outcome. A more vigorous swirl will generate significantly more foam than a gentle swirl. To produce consistent responses, develop a consistent methodology. Teach panelists to gently swirl the glass a few times and then to observe both the size and retention of the head. Panelists should assess both parameters using ordinal scales, with retention measured by a rough estimate of the amount

of time it takes for the foam to collapse. If you choose to track these attributes, you do not need panelists to track retention down to a specific number of seconds, but you should show panelists what qualifies as low, medium, and high ratings within each attribute using commercial examples. Since glass shape and material can affect both head size and retention, make sure that you present your examples in the same type of glassware that panelists will use during sessions.

Visual Assessment Standards

In the modalities of taste and aroma, we often use singular foodstuffs or even individual flavor compounds to train panelists on a given attribute, such as spiking a beer with diacetyl. While we could approach visual assessment in a similar fashion—dyeing a base beer to different shades to achieve precise color standards, or doctoring a beer with tannic acid to produce discrete levels of haze—you have a far better option available to you.

Within the visual realm, you should simply select beers that exemplify the different appearance traits that you want your panelists to learn. For color, teach your panelists the lexical set of color words that you have selected to use and then assemble a palette of differently colored beers to illustrate the scale. For haze levels, you can select examples that illustrate the different points of the spectrum, perhaps using a lager, an unfiltered ale, a hefeweizen, and a highly turbid New England IPA.

While recognition testing is essential in helping panelists master various tastes and aromas, testing attribute recognition for characteristics like color and haze offers fewer benefits. Humans naturally rely on vision more than their other senses, so your panelists will likely display a satisfactory level of visual acuity from the start, assuming they are not visually impaired or color-blind. If you have any visually impaired panelists, simply allow them to skip any assessments related to appearance that they find difficult. Since your other panelists will likely be able to assess the visual traits of your beers with a high degree of accuracy, you will not lose anything by having visually impaired panelists avoid those assessments and you will reduce the potential for anxiety or frustration in sensory sessions.

Aroma Assessment

While inputs from all of our senses contribute to our perception of flavor, aroma drives flavor identity. At the core of a strong sensory practice lies a robust set of aroma sampling techniques. Each technique presented here will help panelists draw certain sets of aromatic compounds from the beer, helping them build a holistic view of a beer's aroma profile piece by piece as they work through each step in turn.

While humans can identify an enormous range of different flavor compounds, we have a harder time detecting differences in aroma intensity (Lawless and Heymann 2010, 38). Our biology itself makes it difficult to accurately assess the level of a given aroma present in a sample. Furthermore, the practice of assigning scale values to the aromas found within a beer forces your panelists to make an additional assessment for each aroma that they encounter, a taxing mental process that will limit the number of samples a panelist can tackle in a given session. Sometimes you may be focused on a specific attribute, for example, trying to determine which of three test batches of your new IPA exhibits the highest level of tropical fruit aroma. In this case, you can ask panelists to specifically scale the level of tropical fruit aroma present or even just ask them to rank the samples based on the intensity of that characteristic. However, in descriptive panels, you should only ask panelists to note which aromas they encounter without worrying about the levels of each individual aroma. When you analyze the aggregate data from your panel, the most prominent aromas will naturally emerge as the descriptors with the greatest numbers of responses (fig. 7.6 on p. 86 shows an example of this).

Before using any of the following aroma sampling techniques, panelists should first swirl the glass. Swirling the glass helps draw aroma molecules out of the beer, making them easier to smell. Additionally, by carefully regulating the way in which panelists swirl their beer, the simple act of glass swirling can act as a subconscious signal to the brain that you are about to commence focused tasting. In his lessons on panelist training, Dr. Bill Simpson of Cara Technology recommends that panelists swirl at "the speed of a record—45 RPM—in an anticlockwise [counterclockwise] direction."[3] The exact speed and direction does not actually matter; bringing a panelist's focus from the beer to the specific way that they swirl their glass will prevent them from overanalyzing the aromas that they identify during evaluation. Counterintuitively, by entering a sort of meditative trance through ritualistically swirling the glass in a certain way, identification of aroma compounds becomes easier.

Panelists should use the following five techniques in sequence to assess the aroma of each beer. In this order, you coax compounds from the sample in descending order of volatility, beginning with highly volatile, easily accessible flavor compounds. While assessing aromas, I recommend you instruct panelists to close their eyes. Blocking out visual stimuli should reduce the impact of appearance on the perceived aromas; also, limiting the total amount of incoming stimuli will make it easier for panelists' brains to process the aromas of the beer.

- **Distant sniff.** Hold the glass six to eight inches (15–20 cm) below your nose and swirl. Use short sniffs to sample any aromas that you perceive from this distance. In most cases, you will not perceive anything using this method, but if the beer contains any extremely volatile aromatic compounds, such as 3MBT,[4] this technique will coax them from the beer, allowing you to assess them before adaptation sets in.
- **Drive-by sniff.** Swirl the glass and draw it across your face, passing the lip of the glass just beneath your nose. Sample the aroma using short sniffs. Like the distant sniff, this technique also helps detect highly volatile compounds.
- **Short sniff.** Swirl the glass, bring it to your nose, and take one to two short sniffs before moving the glass away from your nose.
- **Long sniff.** Swirl the glass, bring it to your nose, and take a one-second-long sniff.
- **Covered sniff.** Cover the lid of the glass, using either a glass cover (e.g., a petri dish) or the palm of your non-dominant hand, first ensuring that your palm is clean and free of any odors. Swirl for five to ten seconds with the glass covered, bring the glass to your nose, uncover the glass, and take a one-second-long sniff. This technique concentrates aroma compounds in the headspace of the glass, amplifying less-volatile aroma compounds such as organic acids and esters.

These five techniques utilize the orthonasal route of olfaction (p. 23). However, we also have a few techniques at our disposal for boosting retronasal perception. Since retronasal detection of aroma

---

3 Bill Simpson, "5-Day Practical Beer Taster Training Course" (workshop), Chicago, IL, October 12, 2015.
4 3-Methyl-2-butene-1-thiol, the compound responsible for skunky notes in lightstruck beer.

occurs when aroma compounds travel from your oral cavity up to your nasal cavity, each of these techniques involves first taking some beer into your mouth. Train panelists to use one of the retronasal techniques discussed below to complete their assessment of the aroma of a beer, focusing their attention solely on aromatic cues before switching gears to assess the tastes present in the beer.

While retronasal olfaction occurs naturally each time you taste a food or beverage, each of the following techniques will help heighten or amplify your perception of retronasal aroma. I encourage you to teach your panelists all three of these techniques, allowing them to select whichever one they find works best. While each technique differs slightly in its execution, they all share the same goal. You are essentially trying to volatilize aroma compounds from the beer to saturate the air inside of your mouth, after which you will push this aroma-rich air out through your nose, briefly inducing an intensity spike in your aroma perception of the beer. Like the covered sniff, these techniques work especially well to boost less volatile aroma compounds, aided in part by the warmth of your mouth increasing their volatility. Additionally, detection of some compounds relies on the increase in pH that occurs when beer enters your mouth and interacts with your saliva. One compound in particular—tetrahydropyridine (THP)—typically only presents retronasally (Snowdon et al. 2006, 6466).

The first method involves aspirating the sample, a technique most commonly known from the world of wine tasting. Personally, I find this technique the most effective of the three. Begin by taking a sip of the beer. Hold the beer in the front of your mouth, purse your lips, and slowly draw air into your mouth, pulling it across the sample. The movement of air across the sample (also known as aspiration of the sample) will typically create a gentle bubbling noise. After drawing in air for three to five seconds, close your lips and breathe out somewhat forcefully through your nose. You want to push the air in your mouth out through your nose, so breath out with some force, but not so hard that you end up blowing your nose (that's gross). If performed correctly, you should experience a burst of aroma as you breathe out.

A word of caution on the aspiration technique—beer, unlike most wines, contains carbonation. When you aspirate the sample, do so gently. As you draw air across the sample, some of the carbonation will escape, causing the beer to foam slightly. If you draw air vigorously across the sample and then breathe out through your nose, you will end up shooting foam out through your nose—a painful and embarrassing fate I am sure you would rather avoid.

To perform the second technique, take a small amount of sample into your mouth and move the sample around your mouth for five to ten seconds to allow the sample to warm slightly. Swallow the sample and then, with your mouth closed, breathe out hard through your nose. Once again, you should perceive a spike in the intensity of retronasal aroma as you breathe out.

The third technique follows the same steps of the second, except that you hold your nose for the entirety of the time that you taste the sample. Begin by pinching your nose and then take a sip of the beer, once again moving the sample around your mouth for five to ten seconds. Swallow the sample and then release your hold on your nose, breathing out through your nose once you have released your grip.

Aroma Assessment Standards

In the realm of aroma training, you have two key groups of tools for training panelists: grocery standards and attribute standards. Grocery standards employ familiar, composite flavors, such as white bread or grapefruit, whereas attribute standards use preparations of individual flavor compounds, such as diacetyl or acetaldehyde. Both groups have their place within flavor training, but you should teach panelists to distinguish between the two.

Grocery standards can help you build a standard, shared lexicon among your panelists. This lexicon becomes the foundation of their tasting vocabulary, useful in building brand descriptions and performing descriptive tastings. To assemble a list of potential standards to use for training your panel, begin by tasting your beers alongside the Beer Flavor Map (see inside front cover). Select standards from the list that reflect the most commonly encountered flavors and use these standards to firm up your panelists' ability to consistently identify those flavors. After exposing your panelists to a group of different grocery standards, you can then perform recognition tests, much as you would with specific attributes (see p. 48 for a description of recognition training and testing).

PREPARING YOUR OWN SPIKES

While preparing your own flavor spikes might save you money compared with purchasing premade spikes, the process entails significant start-up costs and carries a potential safety risk for the person preparing the spikes and for your panelists. For most small breweries, purchasing a limited number of flavor standards offers an easier and more effective route. First and foremost, in order to prepare your own standards, you should have a trained lab technician on staff, or at a minimum someone trained in lab benchwork. The actual act of preparing spikes requires precise measurement of miniscule amounts of the flavorant followed by careful serial dilutions, and these should only be performed by someone with the proper training or skill set.

In terms of equipment, you will first need to install a fume hood to ensure safe preparation of samples. For measuring solid reagents you will need a high-precision scale, while for measuring liquids you will need a volumetric pipette, a calibrated micropipette, or a Hamilton syringe. In many cases, a Hamilton syringe works best due to the high volatility of most pure aroma compounds. For serial dilutions you will want to have a few volumetric flasks, probably ranging between 5 and 200 mL depending on the number of spikes you want to produce at a time. Lastly, you should secure some glass ampoules for storing your finished spikes.

When purchasing a flavorant (e.g., acetaldehyde), make sure to get food-grade chemicals. Consult the safety data sheet (SDS) for each compound to ensure you are handling them properly, because even food-grade chemicals can be dangerous in their pure form. Due to the potentially hazardous nature of these compounds in their pure form, the equipment you use to prepare spikes should not be used for any other purposes.

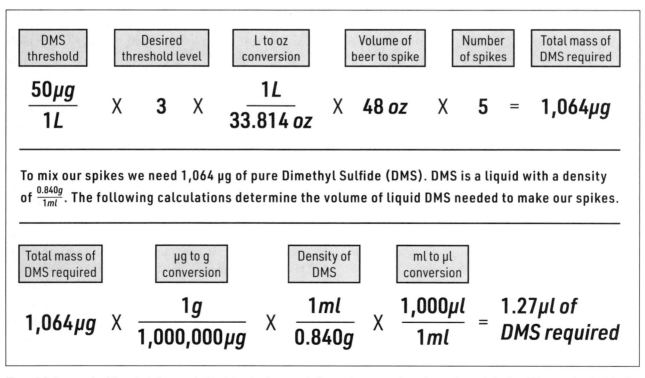

Figure 4.2. An example of the calculations required to determine the amount of reagent necessary to produce a given set of spikes. This example calculates the amount of pure DMS necessary to produce five 1 mL spikes that each yield a three-times-threshold level of DMS when added to 48 oz. of beer (DMS threshold is 50 µg/L). Abbreviations: DMS, dimethyl sulfide; L, liter; ml, milliliter; µl, microliter; oz, ounce; g, gram; µg, microgram.

When preparing a set of spikes, you first need to determine the amount of reagent to use. This will vary based on the threshold of that reagent, the volume of beer that you will spike, the final concentration you want in each spike, and the number of spikes that you want to make. An example of these calculations is shown in figure 4.2. To simplify these calculations, the American Society of Brewing Chemists (ASBC) has a flavor standard spiking calculator available to members that automatically generates the amount of flavor compound needed depending on how much beer you will spike and at what level. Most of the common odorants present in beer have limited water solubility, so you will typically have to use ethanol as a solvent for your spikes. Although you can purchase 95% food-grade ethanol from a chemical supplier, it is significantly more cost-effective to purchase a highly rectified spirit like Everclear, which should not impact the quality of your spikes.

To prepare your spikes, begin by selecting an appropriately sized volumetric flask based on the number of spikes you intend to produce. Following the example laid out in figure 4.2, we would begin with a 5 mL volumetric flask. Add about half of the total desired volume of your solvent (in this example, 95% ethanol); this prefill helps avoid loss when working with highly volatile solutes. Carefully measure the appropriate amount of solute (1.27 μL of DMS) and add it to the volumetric flask. Continue adding solvent to the flask until the liquid level nearly reaches the line marking on the flask. Finish by adding solvent in a dropwise fashion using a pipette until the meniscus of the liquid reaches the line on the volumetric flask. Cap the volumetric flask and gently mix the contents to distribute the solute within the solvent. Lastly, use a volumetric pipette to transfer 1 mL of the spike solution to each of five ampoules. Seal the ampoules and store for use at a later time. Make sure to store your spikes under refrigeration because many flavor compounds degrade over time.

Specific flavor attributes allow you to train panelists on individual flavor compounds by adding single-compound spikes to beer. This type of training can help your panelists identify specific flavor attributes with a high degree of accuracy, but it comes at a cost—prepared flavor spikes typically carry a hefty price tag. The largest companies currently selling flavor standards include Aroxa™, FlavorActiV™, and Siebel.

If you cannot stomach the cost of preprepared flavor standards, you can prepare standards yourself, purchasing food-grade flavor compounds and diluting them to the proper concentration within the lab. This process will cost you both in terms of lab equipment and time, but if you already have a full-time lab technician or sensory technician then preparing your own spikes can save you money in the long run. However, if you do not have a trained technician with experience in precise measurement and serial dilution you should avoid this route as you will likely end up producing unreliable spikes. In the end, the convenience, shelf-stability, and high quality of preprepared flavor spikes lead many companies to use them in spite of the cost. In most cases, I recommend focusing on a small subset of essential attributes, purchasing those from a supplier of flavor standards as necessary.

While grocery standards have their uses you should never substitute them in place of a specific compound when training panelists on a given attribute. In an effort to save money, some trainers opt to use stand-ins for various flavors, such as butter extract for diacetyl or the juice from a can of sweet corn for dimethyl sulfide (DMS). While these grocery standards do mimic the most common flavor descriptors used for these compounds, they do not actually represent the flavor of the compound. The substitutes may or may not actually contain the compound in question, and they also contain other flavor compounds that can draw the focus of your tasters. By training panelists to think of DMS as identical to sweet corn juice, you may end up training them on compounds that will not actually show up in beer. This can become even more problematic with a panelist who is either anosmic to the compound in question or perceives the compound as some flavor other than the most commonly used descriptors.

To give a personal anecdote, I tend to perceive the aroma of butyric acid as overripe papaya or cheddar-cheese popcorn, whereas most literature lists descriptors of baby vomit, baby diaper, or rancid butter. If I tried to find butyric acid in beer based on written descriptions (or, god forbid, tried to train on the compound by intentionally smelling baby vomit), I would never find it, since those descriptors do not match up with my personal perception of that flavor. Indeed, I did not realize that I had in fact encountered butyric acid until after training with pure compound spikes of the flavor. Following that training, I was able to identify the way that I perceived the flavor in beer and can now reliably identify the compound when present.

Recognition Training and Testing

Recognition training offers one of the best ways to train panelists on aromas, whether testing specific attributes or grocery standards. It also happens to work quite well for training on specific tastes. This method of training works best when your samples share identical appearance attributes so panelists cannot distinguish between samples using visual inspection. With specific attribute training, you can serve samples in clear glassware because the spiked samples will all look the same. Training and testing on grocery standards requires some slight modifications; you should use an opaque vessel for presentation of each standard, such as a film canister or a red Solo cup.

Begin by presenting panelists with a group of samples, typically anywhere from five to ten different attributes or aromas. Address each attribute one at a time and encourage panelists to smell and taste the sample while you talk about it. As you discuss a given attribute or aroma, you can provide additional information, such as how it forms in beer or what ingredients tend to drive the presence of that aroma. As you discuss the sample, emphasize the name of the aroma or attribute through repetition, never referring to the flavor as "it," always saying the name that you intend for panelists to use. (For example, "*Diacetyl* is a biproduct of fermentation. Most people describe *diacetyl* as buttery, or butterscotch. *Diacetyl* can also result from bacterial fermentation or dirty draught lines.") Our brains link recognition of aromas to memories or emotions, but we struggle to describe those same flavor impressions using language (Shepherd 2012, 211). This repetition of language during training will help your panelists' brains build connections between a given sensory impression and the word that we assign to describe that perception. Each time a panelist hears the name of a flavor while interacting with a sample, you subconsciously reinforce that association within their brain.

After going through each sample individually, ask the panelists to briefly leave the room. Wearing gloves, scramble their samples, using a predetermined pattern so that all panelists' samples are reordered in the same way (fig. 4.3). Invite the panelists back into the room and ask them to identify each sample. By forcing the panelists to recall the sensory impressions of the attributes or aromas that they just learned, you will help cement the identity of these attributes in their brains, increasing the chance of successful identification when the panelists encounter these aromas or attributes in the future.

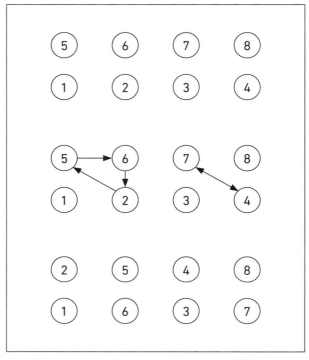

Figure 4.3. An example of a scramble pattern for eight samples. You should vary the patterns you use each time you scramble samples. Increase the complexity of your scramble patterns if you feel that panelists are catching on—the goal is to force panelists to rely on their senses to complete the test.

Tasting Assessment

While a taster holds a sample in their mouth, they can assess at least three different sensory modalities: retronasal aroma, mouthfeel, and taste. However, when a panelist approaches a sample with the goal of specifically assessing its taste then they should attempt to ignore the other inputs as much as possible. When tasting beer, panelists should primarily focus on sweetness, bitterness, and sourness, using ordinal measurements to report the level of each taste. Salinity or saltiness plays a role in some styles as well; if you brew a style featuring notable salinity, panelists should assess levels of this taste as well. While umami can show up in beer due to autolysis or through the addition of unusual ingredients (e.g., bacon), such a situation usually offers other, more telling cues, and your panelists should not need to assess levels of umami taste.

Compared with aroma, tasting beer does not entail much in the way of unique or specific techniques. As we saw in chapter 2 (p. 15), the tongue map myth does not stand up to reality. So, rather than focusing on any specific area of your tongue, you should move the beer around your mouth while tasting to activate as many taste receptors as possible.

When specifically assessing tastes, panelists should both close their eyes and plug their nose. As with aroma assessment, closing your eyes helps to limit incoming visual stimuli. Plugging your nose similarly limits incoming aroma stimuli, but more importantly helps prevent cross-modal interactions between taste and aroma in the brain. Aroma can strongly influence our perceptions of taste, particularly when it comes to sweetness. Smelling notes of caramel, brown sugar, figs, bananas, and other typically sugar-laden foodstuffs augments the perceived sweetness of the beer. Being aware of these effects does not prevent aroma from altering taste perception—the only way to limit the influence of aroma is to block aroma perception through mechanical means by holding your nose while tasting.

Bitterness bears special consideration, both due to its importance as a defining attribute in many beer styles as well as certain peculiarities in the way we perceive hop bitterness. Some bitter compounds produce an instantaneous aversive reaction, an urgent warning from our primitive, survival-oriented brain to expel the substance from our mouth. Hop bitterness yields a softer effect, growing in intensity as it sits on the palate. When attempting to evaluate the level of bitterness in a sample, instruct panelists to take a sip of the beer, move the beer around their mouths, swallow, and then pay attention to how the bitterness level evolves after they have swallowed. Oftentimes, bitterness will crescendo after swallowing and will peak between 15 and 30 seconds after the beer has left the mouth. Carefully noting the point of maximal bitterness will help panelists refine their rating of this taste.

SWALLOW OR SPIT?

In the sensory field, most consumer products are assessed in a way that allows the panelist to avoid actually consuming the food or beverage, typically by spitting the item out after holding it in their mouth for a period of time (Lawless and Heymann 2010, 66). Reasons for spitting differ depending on the product category. Panelists will often spit out calorie-dense products like oils and cheeses to avoid consuming excessive or undesired calories. Tasters of spirits and wine often spit out samples following assessment to avoid intoxication.

Within beer judging and assessment circles, spitting out beer is often viewed negatively. One justification asserts that, due to the importance of bitterness in beer, we must swallow each sample to fully activate bitterness receptors at the back of the palate. However, as we now know, we perceive bitterness (along with the other tastes) all over the tongue. If you spit out beer when sampling, it is true that you will miss activation of taste receptors located in the back of the throat, but you will still be able to produce accurate assessments of the beer's taste. If a panelist prefers to spit out their samples this should not preclude them from participation in panel. However, panelists should make a decision up front to either always spit or always swallow. As with many other facets of sensory evaluation, consistent application of techniques leads to consistent results.

Taste Assessment Standards

While most panels will not perform extensive taste training, you should expose panelists to the basic tastes at least once to help them learn to recognize each taste individually. This helps head off any issues with sour-bitter confusion, which is when a taster confuses acidity for bitterness, or vice versa, and is a relatively common phenomenon. Tasters exhibiting this confusion tend to label sour stimuli as bitter more often than the reverse, with one study showing that 13.3% of respondents labeled a citric acid solution as bitter (O'Mahony et al. 1979, 301). With proper training

Table 4.1 **Preparation of samples for taste training**

Taste	Additive	Concentration
Sweet	Sucrose (table sugar)	5.0 g/L
Salty	Sodium chloride (table salt)	0.5 g/L
Sour	Citric acid	0.3 g/L
Bitter	Iso-alpha acid[a]	Depends on supplier[b]
Umami	MSG (monosodium glutamate)[c]	0.75 g/L

Source: Bennett Thompson (Half Acre Brewing Company), interview, August 9, 2019.

[a] I recommend using iso-alpha acid as it mimics the bitterness present in beer; you should be able to secure iso-alpha acid extract from your hop supplier. However, if you do not want to go to the trouble, you can use crushed caffeine pills to simulate bitter taste. A good starting point is 800 mg/L, but take care with your sample sizes so that panelists' caffeine intake is not excessive.

[b] The concentration of the extract will determine how much you use; aim for 10 IBUs and adjust up or down depending on how that tastes.

[c] MSG can be found as the product Accent Flavor Enhancer in most grocery stores or online.

your panelists can learn to accurately identify each of the basic tastes, which will help them appropriately measure each taste when evaluating beer.

Compared to aroma standards, taste standards are significantly easier to acquire and to use. Most taste standards can be purchased either in grocery stores or online. If desired, you can begin by training panelists on the tastes presented in water before graduating to beer. You can use the values listed in table 4.1 as a starting point for preparing water samples, though note that these amounts are based on empirical data. Taste each sample following preparation and adjust the amounts as necessary.

Use recognition methods with these water-based taste samples to train and test your panelists. Once your panelists can comfortably identify the basic tastes then you can try spiking in beer, which may help train panelists on how the tastes can present differently within a beer matrix. However, many panel leaders find one exposure to water-based training sufficient. On a practical note if adding these substances to beer, begin by dissolving the solid in a small amount of water and then add that water solution to the beer. If you attempt to add the substance directly to beer, you will knock most of the carbonation out of solution while trying to get the solid to dissolve.

Mouthfeel Assessment

When assessing mouthfeel, as with taste, panelists should attempt to limit competing stimuli from other modalities using physical techniques like closing their eyes and holding their noses. You cannot mechanically inhibit taste sensations when assessing mouthfeel (just as you cannot mechanically inhibit mouthfeel sensations when assessing taste) but blocking out aroma and visual stimuli will greatly improve a panelist's ability to focus.

Like appearance, mouthfeel encompasses a variety of different individual attributes. Panelists should always evaluate body and carbonation level when assessing mouthfeel, as these two characteristics play an important role in all types of beer. If present, astringency and alcohol warmth can greatly affect the overall flavor perception of a beer, so you should encourage panelists to comment on each of these attributes as well, noting a level of "none" if the characteristic does not appear. Any additional mouthfeel characteristics that you ask panelists to assess should stem from the types of beers you make. For example, if you brew a beer with chili peppers you will certainly want panelists to evaluate the level of capsaicin or spicy heat present in that beer.

Body

Panelists should describe body using an ordinal scale. To help panelists gauge differing levels of body, instruct them to take a sip of the beer and then push the beer around their mouths using their cheeks in a sort of "swishing" pattern. As the beer moves through the mouth, encourage panelists to pay attention to both the resistance to flow and the residual viscosity of the beer on the surface of the mouth. A word of caution though—vigorous swishing of the beer will cause carbonation to break out of solution, which will both fill the mouth and make accurate perception of body difficult. Sometimes, high levels of carbonation can produce a sensation of lightness as a result of the foam that forms in the mouth. To achieve a more accurate measure of body, you can instruct panelists to swish the beer around their mouths to knock the carbonation out of solution, open their mouths slightly or breathe out through their noses to allow the built up carbon dioxide to escape, and then try moving the degassed beer around their mouths to assess the body of the flat beer. In many cases, panelists will not need to go to these lengths to assess the body of a beer, but this technique can help mitigate the confounding effects of carbonation, particularly in highly carbonated beers.

Carbonation

Assessment of carbonation can cover a number of different elements, including overall carbonation level, bubble size, and irritation quality (whether the carbonation yields a soft, gentle foam or a sharp, biting sensation on the palate.) Of these elements, overall carbonation level offers the most valuable data and should be measured using an ordinal scale.

To assess carbonation level, instruct panelists to swish the beer in their mouths as with the body assessment, paying attention to how much foam builds up. In beers with low carbonation or nitrogenated beers, the beer will barely froth at all, whereas beers with high carbonation will produce a large volume of foam that fills the mouth. Assessing the beer in this manner also allows you to assess irritation quality, as the foam produced may present as soft and creamy or sharp and stinging.

Alcohol Warmth

Alcohol warmth in beer stems from the presence of elevated levels of ethanol and fusel alcohols. We perceive this warmth most intensely on the tongue and the back of the throat. Panelists should assess alcohol warmth using an ordinal scale, with a category for "none" for beers that show no warming sensation. In beers with a high alcohol content, if the warming sensation takes on an unpleasant, burning quality we commonly describe them as "hot." The exact level at which alcohol warmth begins to appear depends in part on other beer characteristics, but, at least anecdotally, most beers below 7% ABV will not show any perceivable alcohol warmth.

To assess alcohol warmth in a sample, panelists should first take a sip of the beer and move it around their mouths. After the beer leaves the mouth (either by swallowing or spitting) the panelist should then breathe out through their mouth. Air moving across our palate will typically increase the perception of alcohol warmth, causing a brief spike in intensity. Using this technique will allow panelists to perceive alcohol warmth even if only present at a very low level.

Astringency

Astringency refers to the puckering or drying sensation on the tongue that occurs when tasting a beverage with high levels of tannins or polyphenols. Since bitterness and astringency often occur together, you should specifically train panelists on the tactile sensation of astringency to help them distinguish between these two attributes. Panelists should measure astringency using an ordinal scale, with a category for "none" for beers that exhibit no astringency.

To assess the astringency level of a sample, take a sip of beer, move it around your mouth, and then swallow or spit out the beer. After the beer leaves your mouth, note any drying sensations that occur on the surface of your tongue, cheeks, or gums. Astringency typically takes a bit of time to register on the palate, usually peaking in intensity five to ten seconds after taking a sip, so when evaluating astringency instruct panelists to wait until at least ten seconds after the beer leaves their mouths before passing a final judgment.

Astringency, unlike most sensory signals, tends to increase in intensity with repeated stimulation rather than showing any sort of adaptation effects. This effect is stronger the shorter the time between sips of beer. If a panelist takes a series of sips in rapid succession their perception of astringency will skew high. However, simply waiting 40 to 60 seconds between sips should reset the palate and the perceived level of astringency should be relatively constant between sips spaced at this interval (Lawless and Heymann 2010, 46).

Metallic

Metallic notes in beer can stem from the presence of metal ions or from "false metallic" aroma compounds resulting from oxidation. False metallic compounds appear during aroma evaluation and should be recorded alongside other aromatic elements of the beer. The techniques presented here focus on how to assess beer for the presence of metal ions. While panelists could measure metallic character using an ordinal scale, any level of metallic character usually indicates an issue, so you can simply have panelists check for the presence or absence of this trait.

When tasting a beer containing metal ions, you will often perceive an assertive metallic aroma retronasally as part of the aftertaste, driven by formation of metallic aroma compounds in your throat. Consequently, using standard retronasal techniques can aid in identification. However, the unique properties of metal ions present us with a few additional techniques that we can use to confirm our suspicions. The first is an old brewing trick, in which you dip your finger into your glass and then rub the beer on the back of your hand. After about five seconds smell the back of your hand—if the beer contains metal ions your hand will give off a metallic aroma, a result of metal ions interacting with the fatty acids present on your skin and forming odor-active compounds that smell like metal (Glindemann et al. 2006, 7006).

The second trick also involves physically touching your beer. This time, swirl the beer to generate some foam, dip your finger in the foam, and then taste the foam off of your finger. If the beer contains metal ions the foam will exude a strong metallic flavor. Iso-alpha acids play a structural role in beer foam and also have a high affinity for metal ions. When metal ions are present in beer, they selectively migrate into the foam, leading to a higher concentration of metal ions present in the foam than in the liquid beer. In a case where you detect metal character in a beer at a low level, sampling the foam can offer confirmation because the foam will typically taste more intensely of metal than the liquid.

Capsaicin

You will only need to assess capsaicin levels in beers made with chili peppers or some sort of related additive. Panelists should measure capsaicin levels using an ordinal scale. Due to a similar mechanism of action, evaluation of capsaicin heat uses the same techniques as assessment of alcohol warmth. Take a sip of the beer and move it around your mouth and then spit out or swallow the beer. After the beer leaves your mouth, breathe out. As with alcohol warmth, breathing out will cause the intensity of capsaicin heat to spike, allowing for more accurate assessment of the level of heat, particularly when present at low levels.

Capsaicin acts on the palate in a variety of interesting ways over time. First, the effect of capsaicin is unique in its duration, in that one exposure to capsaicin can result in sensations lasting ten minutes or longer (in some extreme cases, MUCH longer). Furthermore, repeated exposures to capsaicin have different effects depending upon the time elapsed between each exposure. Rapid sequential exposures to capsaicin will cause the intensity of perceived heat to build with each additional taste. However, if the taster takes a brief rest following the initial exposure they will typically experience some amount of desensitization to additional exposures, at least for a period of time. Lastly, studies have shown that regular exposure to capsaicin (such as eating a diet rich in chilis) leads to chronic desensitization and a higher overall threshold for capsaicin heat (Lawless and Heymann 2010, 43). Due to all of these unique temporal effects, if you do need to assess for capsaicin heat you should implement strict assessment protocols. For example, for beers containing capsaicin heat, you can ask that panelists always assess the level of heat present the first time they take a sip of the beer to avoid the impact of any temporal effects.

Other Sensations

In addition to all of these individual mouthfeel attributes, beer can also display a wide variety of different textural sensations. Both highly carbonated styles (e.g., weissbier) and nitrogenated beers exhibit unique foam textures. Beers brewed with oats often feature an oily or creamy texture. With most of these textural elements, you will want panelists to note the presence or absence of these traits, rather than trying to pin down the degree of, say, creaminess or oiliness present. In such cases, if you want to track these traits, it may make more sense to use a CATA approach rather than individual ordinal scales for each possible texture.

Mouthfeel Assessment Standards

If training panelists on mouthfeel sensations, you should present examples of each mouthfeel attribute individually. While spiking or otherwise doctoring samples works best to train panelists on aroma and taste attributes, you should demonstrate most of these mouthfeel characteristics using standard commercial examples that illustrate each trait.

Body

To train on body, serve panelists different beers covering the spectrum from light and watery to full and chewy. I would suggest using a range of beer styles to map body, such as an American light lager for low body, an American IPA or American brown ale for medium body, and an imperial stout or barleywine for high body.

Carbonation

While you could go to the trouble of carbonating samples of a single beer to different levels, you should be able to train panelists to accurately recognize different levels of carbonation simply using beers from your own brewery or from the market. You should not have trouble finding examples of beer with a medium level of carbonation (~2.5 volumes of CO_2) or beer with a high level of carbonation (3–4 volumes of CO_2). Most American styles of beer—American blonde ale, American pale ale, American IPA, etc.—present a medium level of carbonation, whereas most classic Belgian styles—dubbel, tripel, saison—or German weissbier will exhibit higher levels of carbonation. To simulate lower carbonation levels, begin with a normally carbonated beer and pour it back and forth between two glasses or pitchers to knock some of the carbon dioxide out of the beer. If you produce any nitrogenated beers you should show panelists what to expect in terms of both carbonation level and other unique textures found in this type of product. Lastly, if you think it useful, you can also have panelists taste carbonated water, which likely has a higher carbonation level than you will ever find in beer. In particular, noting the quality of the bubbles in carbonated water will give panelists a clear idea of what sharp, biting carbonation feels like.

Alcohol Warmth

With alcohol warmth, you want panelists to become familiar with the physical sensations caused by alcohol to help them separate that feeling from other potentially confounding sensations. To train on alcohol warmth you can either select a high-alcohol beer or create a dilute solution of a spirit, in which case whiskey tends to work best. As a grain-based spirit, whiskey flavors have some overlap with beer. If training with a spirit, first dilute it to a strength of between 10 and 20 percent ABV before serving it to panelists.

Astringency

Even if you do not ask panelists to track astringency in their mouthfeel assessments, you should train them on the tactile sensation of astringency to help them distinguish between astringent mouthfeel and bitter taste. Preparing an oversteeped pot of tea offers an easy and accessible way to help panelists isolate the way astringency feels on the palate. Alternatively, low levels of dissolved alum (2–3 g/L in water or beer) can also stimulate an astringent response.[5]

Metallic

Metallic flavor should only be added to beer using premade spikes. While creative approaches to problem solving can serve you well in some areas of sensory work, I advise against trying to find a novel way to infuse metallic flavor into your beers.

Capsaicin

Most panelists probably have enough experience with capsaicin heat from consuming food with varying degrees of spicy heat. If you want to train on this attribute within a liquid matrix, however, you can mix up samples of either water or beer with varying levels of cayenne pepper as a fast and easy way to present discrete levels of capsaicin heat.

---

[5] Bennett Thompson (Half Acre Brewing Company), interview with author, August 9, 2019.

5
BEER FLAVOR

Compared with most other beverages, beer presents an enormous array of different potential flavors. From the walnut and cocoa notes of your favorite American brown ale to the ripe mango and pineapple aromatics of a New England IPA or the banana and clove fermentation profile of a classic German weissbier, brewers paint a kaleidoscope of flavor using standard beer ingredients. Step a bit further afield and you will find the bacony smoke of traditional rauchbier, the tart apricot and barnyard funk of a Belgian gueuze, or the brown sugar, vanilla, and marshmallow of a bourbon barrel–aged imperial stout, all while remaining within the lines of accepted, definable styles. However, today's brewers know no bounds when it comes to the use of unique and imaginative ingredients, from fruit and spices to chili peppers and shellfish. The modern beer landscape offers virtually endless variety.

In building your sensory program, one of your first tasks will be to establish a flavor lexicon that your panelists will draw from when evaluating your beers. The descriptors that you and your panelists choose will form the foundation of your brand profiles, which you will use to assess outgoing beer at each and every product release panel. While this may all sound daunting, much of the heavy lifting has already been done. The Beer Flavor Map (see sidebar and inside cover of the book) contains a wide-ranging set of beer-specific flavor descriptors and should serve as the starting point for your panel's flavor lexicon. The map is quite robust and many panels will likely find the flavors listed there sufficient. However, if you want to go beyond the flavor map or simply adjust some of the terms, work with your panelists to shape a custom lexicon to cover the most important flavors found in your beers. Each panelist brings with them a lifetime of diverse flavor experiences. As the panel leader, you serve as their guide, helping the group reach consensus when deciding how to name different flavors. Even with very little training or experience tasting together, groups can often align on the key flavors present in a beer.

In contrast with other product categories, sensory work in beer often places greater emphasis on finding flaws than noting positive traits. In the worlds of wine or coffee or cheese, sensory specialists usually look for "on" flavors—the positive characteristics that define a product—rather than product defects. For whatever reason, beer tasters often search first for faults before turning their attention to whether the beer tastes as it should. This approach seems somewhat contrary to logic. First off, our brains tend to find what they are looking for—if we begin by searching for an off-flavor, we increase the likelihood of finding evidence of one, whether real or imagined. Secondly, when I think of the key characteristics of my favorite IPA, the first thing that comes to mind is not its lack of diacetyl and DMS but rather its hop bouquet of passion fruit, mango, and tangerine, perched atop a foundation of white-bread malt flavor and assertive bitterness. To ensure that your brand profiles feature positive flavor descriptors, you will want to teach panelists to recognize these positive characteristics as opposed to focusing flavor training exclusively on finding brewing flaws or other negative flavor attributes.

BEER FLAVOR WHEEL AND BEER FLAVOR MAP

In the 1970s, a group of sensory scientists led by Morten Meilgaard (a highly regarded scientist within the field) developed the Beer Flavor Wheel with input from the American Society of Brewing Chemists (ASBC), the European Brewery Convention (EBC), and the Master Brewers Association of the Americas (MBAA). They aimed to develop a universal lexicon of beer terminology, hoping that brewers, panel leaders, ingredient suppliers, and other members of the industry would use it to communicate beer flavor descriptions in a standardized way. Meilgaard recognized that the wheel was a product of its time, its components based on contemporary knowledge of beer flavor. The original publication stipulated that a new working group should revisit and update the flavor wheel within a few years to factor in new research and understanding of beer flavor—this way the wheel would remain relevant (Meilgaard, Dalgliesh, and Clapperton 1979, 47). However, to date, the flavor wheel has never been updated.

In 2016, Lindsay Barr and Dr. Nicole Garneau released the Beer Flavor Map, a sort of spiritual successor to the Beer Flavor Wheel. Not only has our understanding of beer flavor greatly improved since the 1970s, but the types of beer and the array of flavors found in beer today have grown significantly since Meilgaard first introduced the original wheel. The Beer Flavor Map captures a much wider scope of potential flavors and serves as an excellent foundation for building a lexical set of beer flavor descriptors. In the few years since its release, the map has been embraced by the broader beer sensory community, seeing use in sensory programs across the nation. In the past, judges at the Great American Beer Festival® and the World Beer Cup® could find a copy of the Beer Flavor Wheel included in their judging packets as a lexical aid; today, judges receive copies of the Beer Flavor Map. To aid in training your panelists, you can find a reproduction of the Beer Flavor Map printed on the inside cover of this text.

Some people hold the dangerous misconception that you cannot build an effective sensory program unless your panelists are trained and validated on a wide variety of specific flavor attributes. I have had conversations with brewers who had yet to begin setting up a sensory program because they did not have the time and resources necessary to train panelists to recognize a certain set of attributes. This mentality stems from an obsession with avoiding flaws and ignores the fact that new panels should focus primarily on brand training rather than specific flavor training (for more information on brand training, see p. 112). By developing an intimate familiarity with your beers, your panelists will successfully identify any flaws that appear, as those rogue flavors will fall outside of the target brand profile. And while your panelists may not be able to identify a fault by name at this stage, recognition of the presence of an unexpected flavor is still extremely valuable information. Do not allow a lack of attribute training to hold you back from getting your sensory program off the ground.

Later in this chapter, I identify a few attributes that I consider valuable within most breweries. If you train panelists on specific attributes, you want them to internalize their perceptions of each flavor compound so that they can reliably identify those flavors not just at the end of the training session, but in the following weeks, months, and even years. By focusing on a smaller set of key attributes, you allow your panelists to build confidence and acuity with the most important flavors that they will encounter. Over time, you can add additional flavors to your training regimen as you see fit; but remember, each additional flavor that you decide to train on costs you in terms of both time and money. Weigh the decision to tack on an additional flavor carefully and, as always, with an eye toward the attributes most important in the types of beers you produce. As a good litmus test for adding an additional attribute to your training regimen, you should be able to point to a specific situation in which knowledge of that attribute would have improved the panel's data or facilitated better decision-making.

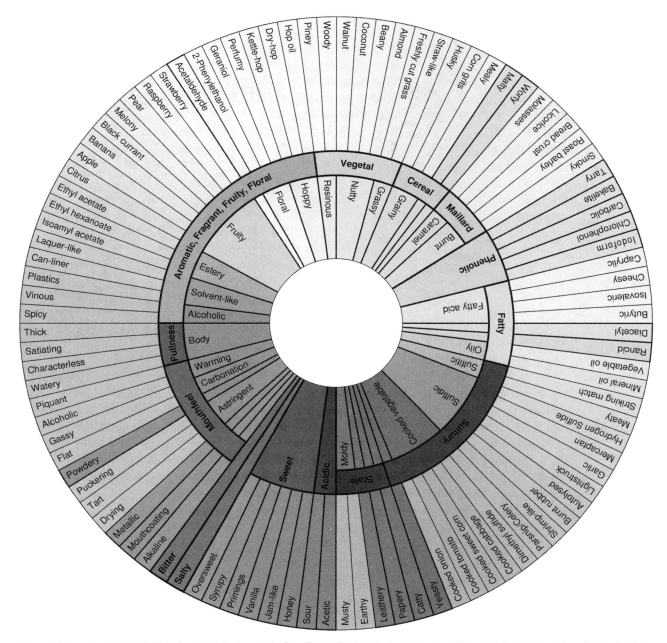

Figure 5.1. Developed in the 1970s by Dr. Morten Meilgaard, the Beer Flavor Wheel marked an important milestone in the path towards understanding beer flavor. Today, it has largely been supplanted by the Beer Flavor Map. Courtesy of beerflavorwheel.com and Abraham Kabakoff.

INGREDIENT FLAVOR DESCRIPTORS

When teaching panelists descriptive vocabulary, doing so through the lens of beer ingredients can help them contextualize flavors by connecting descriptors to the ingredients that they come from. This provides a natural organizational paradigm that helps panelists build cause-and-effect associations, connecting recipe and process details to the samples they taste as they deepen their understanding of your beers. Many of the specific flavors mentioned in this section correspond with descriptors on the Beer Flavor Map. When first tasting through your brands with new panelists, you may find it helpful to have them taste alongside the flavor map, using it as a reference. These flavor descriptors will then form the basis of the target profiles your panelists develop for your brands.

Malt-Derived Flavor

Differences in flavor between types of malt derive from conditions in the kilning phase of the malting process.

DEFINING AN OFF-FLAVOR

Over the years, I have both hosted and attended a number of attribute training sessions billed as "off-flavor training." Offering "flavor training" fails to grab the attention in the same way as the urgent and somewhat alarmist "off-flavor training" does. However, I personally try to avoid using the term off-flavor to describe specific flavor attributes. Context determines the appropriateness of a flavor attribute in a given beer—rare is the flavor that is universally unacceptable. Diacetyl, undesirable in most styles of beers, supports the malt profile in many classic British styles and plays a key role in the classic Czech premium pale lager, Pilsner Urquell. Acetaldehyde, a symptom of young or immature beer in most cases, can add a complimentary fruity note to the profile of an American lager. Even butyric acid—commonly described as baby diaper or baby vomit—can play a positive role due to the tropical fruit notes it evokes at low concentrations in certain beers.

A better definition of an off-flavor comes back to brands and intention. When you design a brand, you do so with certain flavors in mind. An off-flavor denotes any flavor outside of the desired profile of a given brand. For example, in an IPA that should feature aromas of mango and lemongrass, piney or resiny hop notes could be termed an off-flavor. In this definition, an off-flavor is not inherently unpleasant, but instead simply signifies an out-of-place flavor within a given beer.

Figure 5.2. Several different types of malted grain.
© Getty/Christine Pickering Photography

Depending on the moisture content of the malt, kilning temperatures employed, and duration of kilning, maltsters can coax a wide variety of distinct flavors from the grain. We sometimes describe malt flavors through the analogy of "bakery flavors," acknowledging the similar flavors found in baked goods. This relationship makes sense, as the kilning process essentially involves placing malt in a giant oven.

Pale malts yield flavors like wheat flour, bread dough, white bread, and water cracker. As malts grow darker in color, they begin to elicit flavors of toast and bread crust. Continuing on to darker malts, we encounter flavors of nuts, chocolate, espresso, and, eventually, burnt and acrid notes in the case of the darkest of malts. Crystal malts, which are kilned while still rather moist, follow a slightly different trajectory, with pale crystal malts yielding notes of honey and caramel, while darker crystal malts give flavors of toffee and dark or dried fruits like plums, raisins, and figs. Smoked malts—kilned using direct heat from a fire rather than heated air—develop smoky flavors ranging from bacon and ham, to campfire or mesquite barbecue depending on the fuel used to stoke the fire.

Hop-Derived Flavor

With the IPA boom of the last decade, hop flavor has grown into a topic of ever-increasing complexity, as new varieties with novel flavor traits continue to enter the market. Traditionally, hop flavor was categorized by region. German and Czech hops, such as Hallertau, Spalt, Tettnang, and Saaz, produce floral, minty, peppery, and spicy notes. Classic English varieties like Fuggle and East Kent Goldings provide earthy, herbal, and green tea aromas. American hops, such as Cascade, Centennial, and Chinook, give flavors described as pine resin and citrus, often leaning toward grapefruit and orange.

Today, a much wider variety of hop flavors exists, and former geographic distinctions begin to fall apart

Figure 5.3. Hop cones on the bine.
© Getty/Grotmarsel

upon closer examination. Newer American varieties, such as Simcoe, Amarillo, Citra, and Mosaic, yield flavors ranging from tropical fruit like pineapple and mango to dank marijuana notes and the pungent sulfur notes of green onion and garlic. New German varieties like Mandarina Bavaria and Hüll Melon break the traditional mold with their fruit-forward aroma profiles. Australian and New Zealand hops like Galaxy and Nelson Sauvin yield unique flavors of stone fruit, lychee, and white wine grapes.

The range of flavors produced by hops only continues to grow as time goes on, with new entrants like Sabro providing coconut tones. This is an exciting era in terms of the wide world of possible hop characteristics—you should expect that options will only increase in the coming years.

Fermentation-Derived Flavor

Malt and hops take well to composite flavor descriptors of common foodstuffs like caramel or grapefruit, with complex aroma profiles comprised of hundreds of different odor-active compounds. Conversely, many fermentation flavors correlate directly with specific attributes, leading some tasters to cite individual compounds in place of descriptors, for example, isoamyl acetate (banana) or 4-vinyl-guaiacol (clove). However, it is entirely reasonable to stick with general flavor descriptors to cover the different fermentation profiles produced by yeast and other fermentative organisms.

We divide all brewing yeast strains into two main groups based on species: *Saccharomyces cerevisiae*, commonly referred to as ale yeast, and *Saccharomyces pastorianus*, commonly referred to as lager yeast. Of the two, ale yeast strains produce significantly more characterful fermentations. In teaching new tasters the differences between a lager and an ale, we often point to the fact that ale yeast produces esters—a group of fruity, fermentation-derived flavor compounds—while lager yeast does not. However, this is a bit of an oversimplification. A lager yeast, if subjected to the higher fermentation temperatures used to ferment ales, can produce significant levels of esters, and some strains of lager yeast will produce low levels of esters even at the cooler temperatures typically used to ferment lagers (Philliskirk 2012). However, in most circumstances, ale yeast will typically produce higher levels of esters in their fermentations.

Although a lager yeast rarely adds much in the way of additional flavors to the profile of a beer, it will sometimes contribute various sulfury notes, ranging from the minerally, struck match character of sulfur dioxide (SO_2) to the eggy notes of hydrogen sulfide (H_2S). An ale yeast will often make its presence known, potentially producing a variety of distinct fermentation flavors. As such, ale fermentations can lead to a wide array of different estery notes, including banana, pineapple, pear, and apple. Additionally, yeast strains that exhibit the phenolic off-flavor (POF) phenotype can produce a range of phenolic compounds during fermentation, typically yielding spicy flavors like black peppercorn or clove (Boulton and Quain 2001, 522).

The world of fermentation really opens up when we consider organisms other than *Saccharomyces* species. The most commonly used genus of yeast outside of *Saccharomyces* is *Brettanomyces* (also known as "Brett"), which produces flavors ranging from pineapple to barnyard, horse blanket, wet wool, and leather. While most Brett flavors do not sound pleasant in isolation, they build beguiling layers of complexity in certain styles of beer. Producers of acid-driven beers exploit still other fermenters, turning to bacteria such as *Lactobacillus* and *Pediococcus* to sour their beers.

Water-Derived Flavor

Compared with the three other primary beer ingredients, water typically takes a backseat when it comes to discussing flavor. That is not to say that water is unimportant. Water makes up the majority of beer by weight and its chemical composition influences virtually all aspects of the brewing process, from mashing to boiling to fermentation. In general, water tends to affect how the other ingredients come together rather than providing explicit flavors of its own. However, water can carry certain undesirable characteristics, such as chlorine, metallic notes, and earthy flavors. Brewers should evaluate their source water on a daily basis to ensure that it is free from flaws (see p. 138).

Other Flavors

Outside of flavors provided by the core ingredients used to make beer, novel processes and ingredients open up entirely new vistas of possible flavors. Barrel aging can impart numerous wood-derived flavors, such as vanilla or coconut, and flavors associated with the previous resident of the vessel, whether bourbon, tequila, or port. Other ingredients, whether fruits, spices, herbs, or any number of exotic additives, will impart their own novel flavors to beer. The dizzying array of options available to modern brewers could fill another book, and, indeed, some authors have devoted entire texts to exploring the bounds of possible beer flavors. However, we have yet to discuss one very consequential process that invariably occurs in all beers, albeit often against the brewer's wishes: oxidation.

Beer oxidation as an umbrella term describes a host of different chemical reactions that occur in beer as it ages. While different beers will age differently depending on the way they taste when fresh, we can establish some predictable guidelines when it comes to the development of oxidation flavors. In general, the first characteristic to change is hop flavor and aroma, as these compounds break down most readily. Following some loss of hop character, bitterness declines and malt flavor tends to shift, with new flavors emerging depending on the color of the beer. Very pale beers take on honey-like oxidation flavors, dark gold and amber beers show increased levels of caramel and toffee flavors, and dark amber to black beers develop increasing levels of dark and dried fruit flavors like prune, date, fig, and raisin. Waxy, cardboard, and papery notes typically appear as oxidation

progresses. Various beer traits, such as alcohol content and color, can affect the exact amount of time it takes for the oxidation process to run its course, but storage temperature unquestionably plays the primary role, with higher temperatures leading to faster rates of oxidation (Vanderhaegen et al. 2006, 358).

FLAVOR THRESHOLDS

The flavor thresholds reported for each attribute in this chapter come from Cara Technology. These thresholds provide a rough idea of the average threshold for each compound but should not be taken as gospel. Reported thresholds in literature vary widely from source to source, sometimes by as much as ten or a hundred times. Complicating matters, the perceivable threshold for a given compound changes depending on the surrounding matrix—for example, diacetyl spiked into a light American lager would likely exhibit a significantly lower perception threshold value than diacetyl added to an imperial stout. Further confounding these issues, sample pH can alter the perception threshold of certain compounds. For example, organic acids like acetic and isovaleric acid become significantly more flavor active as pH decreases. If all that were not bad enough, perception thresholds also vary at the individual level, influenced not only by genetic factors, but also by behavioral factors such as diet and consumption habits (Meilgaard, Civille, and Carr 2016, 24). So, while the thresholds provided here can give you an idea of the relative differences in aroma activity from one compound to the next, do not give too much weight to the exact numbers themselves.

SPECIFIC ATTRIBUTES

While I want to avoid perpetuating the focus on faults, training panelists on a small set of specific attributes will help them recognize the most commonly found flaws should they appear in your beers. As your program grows, you may wish to train panelists on a wider group of flavor compounds, but do not rush to expand the pool of your attributes. Racing through a series of

Threshold concentrations are typically very low, in the realm of parts per million, billion, or even trillion. Here are some common concentrations, from high to low:

Gram equivalent	Concentration	Units	Parts-per equivalent	Abbreviation
1 mg = 1×10^{-3} g	milligrams per liter	mg/L	parts per million	ppm
1 µg = 1×10^{-6} g	micrograms per liter	µg/L	parts per billion	ppb
1 ng = 1×10^{-9} g	nanograms per liter	ng/L	parts per trillion	ppt

trainings covering 25 to 30 attributes with just one or two exposures to each will likely leave your panelists lost and confused. You will derive far more value from attribute training if you present panelists with a core group of five to eight attributes, offering multiple exposures to each compound to help them hone their recognition abilities. If you are just starting out, pick a group of five or so attributes from the following section, and focus attribute training solely on those flavors. If you are looking for a specific list to start, I recommend diacetyl, acetaldehyde, isovaleric acid, acetic acid, and hydrogen sulfide. Over time you can incorporate more flavors, but always think first about whether an additional attribute will actually improve the data gathered by your panel before presenting it to panelists.

Diacetyl

Diacetyl
Threshold: 10–40 µg/L (ppb)
Common flavor descriptors: butter, butterscotch, movie-theater buttered popcorn

If you only train your panelists on one specific attribute, make it diacetyl. Diacetyl appears in beer more frequently than any of the other attributes presented in this list. Some styles contain moderate amounts of diacetyl as part of their profile. Many English ale styles feature low levels, and the classic Czech premium pale lager Pilsner Urquell has a specification for the amount of diacetyl present. However, in most cases, diacetyl is an undesirable flavor, indicating a flaw or some other issue.

Diacetyl most frequently appears in beer as a by-product of fermentation. During the cell growth phase at the outset of fermentation, yeast produces a compound called α-acetolactate as a building block in amino acid synthesis. Some of this compound invariably leaks out of the yeast cells, slowly undergoing an oxidative transformation to form diacetyl. Following this oxidation step, the yeast will reabsorb diacetyl and reduce it to acetoin and subsequently 2,3-butanediol, both of which provide no contribution to flavor at the concentrations typically found in beer.

While active yeast will reabsorb and reduce diacetyl in an instant, the oxidative conversion of α-acetolactate to diacetyl proceeds at an abysmally slow pace by comparison (Inoue 2008, 58). In a biological or chemical system, we call this the rate-limiting step, as the rate of this reaction determines the rate at which the entire series of reactions can proceed. The problem is that α-acetolactate does not present any flavor cues of its own. As a result, you could assess the profile of a maturing sample searching explicitly for diacetyl and, finding none, assume that maturation had run its course. However, using this technique you will have no way of knowing whether the sample still contains α-acetolactate. If you push beer through the maturation stage too quickly, you may decide to crash the yeast and proceed to package the beer. In the weeks following packaging, any α-acetolactate remaining in the beer will be converted into diacetyl. Even if you carbonate the beer using bottle conditioning, the small amount of yeast present will not be able to counteract the impact of a maturation period cut short. When poured into a glass, those familiar buttery notes will rise to greet the drinker, indicative of the high level of diacetyl present.

Most breweries consider maturation complete when the beer no longer contains α-acetolactate. In large breweries, lab technicians will measure α-acetolactate levels using gas chromatography–mass spectrometry (GCMS), which can measure the chemical constituents of a sample of beer to a high degree of accuracy. But the precision and analytical power of GCMS comes at a cost—you will

DIACETYL FORCE TEST

While qualitative rather than quantitative, the diacetyl force test (sometimes referred to as a VDK force test)* allows brewers to accurately assess whether any α-aacetolactate (the diacetyl precursor) remains in a beer following maturation. The test relies on the fact that the conversion of α-acetolactate to diacetyl—like many chemical processes—occurs at a faster rate at higher temperatures. Subjecting a sample of beer to elevated temperatures rapidly converts any α-acetolactate present into diacetyl. After briefly heating the sample, you can assess the aroma, searching specifically for notes of diacetyl. Use the diacetyl force test on beers that have reached terminal gravity to determine whether maturation is complete.

The following steps outline how to perform a diacetyl force test (White and Zainasheff 2010, 223).

- Prepare a hot water bath between 60°C and 71°C (140–160°F).
- Draw two tank samples of the beer you wish to sample, and cover them with aluminum foil.
- Label one of the samples as "control" and store it at room temperature.
- Place the test sample in the hot water bath for 10–20 minutes.
- Allow the test sample to cool to room temperature using a cold-water bath, refrigeration, or simply time.
- Remove the foil from both samples and compare their aromas.

If your test sample displays a diacetyl aroma, that indicates that some α-acetolactate still remains in the maturing beer. Perform the diacetyl force test daily until the test sample comes back negative for diacetyl aroma. This simple test can save you significant time and headache, preventing you from having to recall a diacetyl-laden beer from the market.

* VDK stands for "vicinal diketone," which describes a group of chemical compounds that includes diacetyl.

have to cough up anywhere between $50,000 and $100,000 for a used unit, with new units running even higher. Many small brewers instead use the diacetyl force test (see sidebar, p. 62) to determine when maturation has finished.

Diacetyl can also result from fermentation by some non-*Saccharomyces* organisms. Many lactic acid bacteria, such as *Lactobacillus* and *Pediococcus*, can produce diacetyl, with *Pediococcus* species in particular producing prodigious amounts. While brewers sometimes use lactic acid bacteria in the production of sour beers, these bacteria can also grow in dirty, infected draught lines (Storgårds 2000, 20). If possible, you should also train taproom staff and salespeople on diacetyl, teaching them to recognize the flavors associated with dirty draught lines. This can pay significant dividends in the long run.

Acetaldehyde

Acetaldehyde
Threshold: 5 mg/L (ppm)
Common flavor descriptors: green apple, pumpkin, latex paint, fresh-cut grass

Acetaldehyde, like diacetyl, also appears in immature beer, and is likewise one of the most frequently encountered attributes in finished beer. The core chemical transformation of fermentation is the conversion of sugars into ethanol and carbon dioxide. However, this process does not occur in one step, but takes place through a concerted series of reactions with many intermediate molecules formed along the way. In the last step of the chain, yeast reduce acetaldehyde to form the final product, ethanol. Thus, actively fermenting or immature beer typically contains a significant amount of unconverted acetaldehyde. While finished beer containing acetaldehyde often indicates premature removal of beer from yeast, high aeration levels, high pitching rate, and elevated fermentation temperature can all exacerbate acetaldehyde formation as well (Stewart and Russell 1998, 52).

Organic Acids

Although, chemically speaking, all of the compounds that fall under this category are acids, only a few of them contribute to sour taste at the concentrations found in beer. Lactic acid constitutes the primary driver of acidity in most sour beers, though acetic acid can play a secondary role as well. Organic acids like citric acid, malic acid, and succinic acid can also contribute to acid taste, particularly when present as a result of the addition of fruit.

Most odor-active acids have perception thresholds in the low parts per million, and so contribute aromas to beer long before they would have any impact on taste. These include the fatty acids isovaleric acid, butyric acid, and caprylic acid, which yield flavors generally considered unpleasant. As a group, organic acids tend to be a bit less volatile than other classes of compounds, so techniques like the covered sniff (p. 44) can help to coax them out of a beer.

Lactic Acid

Threshold: does not have much of an aroma impact, so threshold is typically not reported

Common flavor descriptors: tart, yogurt, dairy

Lactic acid drives the acidity present in sour beers. In explicitly sour beer, brewers will typically ferment the beer with some species of lactic acid bacteria. *Lactobacillus brevis*—a favorite among brewers producing rapidly soured beers with a neutral profile—ferments relatively cleanly, often producing solely lactic acid. *L. brevis* can be used in a traditional fermentation environment, though in recent years it has seen increasing use as a kettle souring agent. To kettle sour a beer, the brewer adds *Lactobacillus* to wort immediately following lautering. The wort is held in the kettle around 38°C (100°F), which induces a rapid lactic acid fermentation, acidifying the beer within 24 to 48 hours (Tonsmeire 2014, 138).

Brewers will also sometimes recruit *Pediococcus* to acidify a beer. *Pediococcus* species often participate in mixed-culture fermentations, working alongside the numerous strains and species of bacteria and yeast that make up these complex cultures. In addition to lactic acid, *Pediococcus* produces tremendous amounts of diacetyl, and should always be paired with an organism that can easily reduce the diacetyl formed (Sparrow 2005, 113). Coincidentally, mixed cultures such as these also often contain some *Brettanomyces* strains, which just so happen to have a high capacity for diacetyl reduction (Cilurzo 2012).

Brewers wishing to add lactic acid without introducing lactic acid bacteria into their breweries do have a few other options at their disposal. Acidulated malt—produced by allowing the naturally occurring lactic acid bacteria found on the grain to grow during malting—is often used in small amounts for pH adjustment, but in larger quantities can contribute enough lactic acid to produce moderate acidity in beer. Similarly, brewers can simply add lactic acid directly to adjust the flavor of their beers. Brewers regularly use lactic acid in the same manner as acidulated malts, using low levels to adjust mash pH. And, while uncommon, I have tasted sour beers acidified solely through the addition of a concentrated lactic acid solution.

These methods describe how lactic acid can arise in beer when brewers desire its presence. But lactic acid can also appear unintentionally, usually through contamination with the same bacteria, namely, *Lactobacillus* or *Pediococcus*. The flavors produced by these bacteria illustrate the utility of our prior definition of an "off"-flavor, that is, whether we consider the bacteria an essential part of the process or a beer spoiler all depends on the context.

Acetic Acid

Threshold: 90 mg/L (ppm)

Common flavor descriptors: vinegar

While many flavor compounds have multiple reference descriptors, acetic acid features but one: vinegar. This association is so strong because the primary aroma of vinegar is, in fact, acetic acid. Acetic acid bacteria, such as *Acetobacter* or *Glucanobacter*, produce acetic acid through oxidization of ethanol. You may have noticed that different varieties of vinegar tend to be associated with specific alcoholic products, for example, red wine vinegar, rice wine vinegar, cider vinegar, champagne vinegar, or sherry vinegar. Traditional manufacturers produce vinegar by first making the corresponding alcoholic product and then allowing *Acetobacter* to convert all of the ethanol present to acetic acid (Adams 2014, 717).

Acetic acid bacteria are obligate aerobes, meaning they require oxygen to live. As a result, acetic acid formation can only occur in the presence of oxygen. In the brewery, acetic acid production most commonly occurs in barrel-aged beer, because wood, being microporous, allows for slow, continual ingress of oxygen. Some styles like Flanders red ale and lambic-style beers do call for low levels of acetic acid, but brewers often aim to limit this flavor in other styles due to its harsh and slightly solventlike nature. To limit acetic acid production, producers of barrel-aged beer use careful filling and sampling techniques to reduce oxygen exposure and will sometimes use cooler temperatures (21°C/70°F or lower) to blunt the growth of *Acetobacter* (Tonsmeire 2014, 171).

Isovaleric Acid
Threshold: 1 mg/L (ppm)
Common flavor descriptors: cheese, parmesan, sweat sock, gym bag

Isovaleric acid most commonly arises through the use of old or oxidized hops. Hops, when exposed to oxygen, begin to undergo oxidative degradation, leading to the accumulation of isovaleric acid. This flavor forms in raw hop material, not in hopped beer as it ages. Consequently, raw hops can be assessed for the presence of isovaleric acid by performing a quick hop rub or hop grind prior to use (see p. 137). Hops that smell of isovaleric acid should not be used, as this flavor will carry over into the finished beer.

Sulfur Compounds

Sulfur compounds cover a wide range of different flavors and tend to exhibit odor activity at very low concentrations, with thresholds ranging from low parts per billion down to low parts per trillion—between 1,000 and 1,000,000 times more odor active than most organic acids or esters. Many sulfur compounds present unpleasant, off-putting aromas, such as the rotten vegetable character associated with dimethyl disulfide (DMDS) or the sewage, hot garbage–like flavor contributed by methanethiol (often termed "mercaptan"). Lightstruck, or skunky, aroma also results from a highly odor-active sulfur compound, 3-methyl-2-butene-1-thiol (3MBT). However, not all sulfur compounds produce unpleasant aromas. A group of hop-derived thiols, which includes 4-methyl-4-mercapto-2-pentanone (4MMP) and 3-mercaptohexyl acetate (3MHA), give hops aromas of gooseberry, guava, black currant, passion fruit, and grapefruit. Though these flavors can sometimes overwhelm due to their extremely low flavor thresholds, many people find these traits positive and enjoyable.

Sulfur-containing compounds are often highly volatile (i.e., they readily become airborne), and we also tend to acclimate to sulfur compounds a bit more rapidly than other classes of compounds. As a result, techniques such as the distant sniff, the drive-by sniff, and the short sniff (p. 44) can help panelists identify sulfur compounds in beer before adaptation sets in.

WHAT'S IN A NAME?

In both beer and wine, 4MMP (4-methyl-4-mercapto-2-pentanone) is sometimes described as "catty." This moniker results from the fact that 4MMP smells quite similar to a phero-mone found in male cat urine (Miyazaki et al. 2006, 1074), the mere thought of which puts some drinkers off of the flavor entirely. Howev-er, 4MMP also shows up in gooseberries, black currants, and passion fruit. As an anecdotal ex-periment, Dr. Bill Simpson trained two different groups of tasters on the flavor compound 4MMP, describing the flavor as black currants to one group and cat pee to the other. Later in the week, when asked to hedonically rate a beer (i.e., rate how much they like the beer) containing 4MMP, the first group rated the beer significantly higher than the second. Although both groups received the same exact beer, the context of their expe-riences differed based on the associations they had been taught regarding this specific flavor.[*] This simple experiment demonstrates the power that flavor descriptions you assign to your beer can have on both panelists and consumers.

[*] Bill Simpson, "5-Day Practical Beer Taster Training Course" (workshop), Chicago, IL, October 14, 2015.

Hydrogen Sulfide
Threshold: 4 µg/L (ppb)
Common flavor descriptors: eggy, rotten eggs

In the course of a standard fermentation, yeast produces significant amounts of hydrogen sulfide (H_2S). However, the carbon dioxide released by a vigorous fermentation will actively strip most H_2S from the beer. As a result, lager beers tend to exhibit higher levels of H_2S compared with ales because the lower temperatures employed in lager fermenta-tions lead to less vigorous fermentations. Elevated sulfate levels in brewing liquor can also increase the amount of H_2S produced (Hampson 2012). Beer from

Burton-on-Trent in England has been described historically as having a peculiar aroma called the "Burton snatch," a combination of elevated levels of H_2S and sulfur dioxide (giving a minerally, struck match note) resulting from the extremely high levels of calcium sulfate found in Burton's natural water supply.

While low levels of H_2S may appear in fresh lagers, high levels often indicate either fermentation issues or contamination.

Dimethyl Sulfide (DMS)

Threshold: 30–50 µg/L (ppb)

Common flavor descriptors: sweet corn, cooked cabbage, tomato paste

Dimethyl sulfide (DMS) proceeds from a precursor found in barley malt, specifically, an amino acid called *S*-methylmethionine (SMM). When exposed to heat, SMM degrades to form DMS. High malt kilning temperatures also degrade SMM, but the heat largely volatilizes the DMS that forms. As a result, very pale malts like two-row, six-row, and Pilsner malt tend to have higher DMS potential compared with other malts. When these malts are used in the brewhouse, an insufficiently vigorous boil or an overly long whirlpool stand (more than 2 hours) prior to cooling can lead to elevated levels of DMS in the finished beer. Another source of DMS is dimethyl sulfoxide (DMSO), which is also found in malt. Yeast can reduce DMSO to DMS, but this pathway is less common and typically accounts for a lower proportion of total DMS (Fix 1999, 34). At low levels, DMS can accentuate the malt traits present in pale styles like Munich helles or cream ale, but at higher levels takes on an unpleasant vegetal characteristic.

Esters

During fermentation, yeast produces esters by joining an alcohol with an organic acid. The naming convention for esters is to name the alcohol followed by the acid, for example, ethyl acetate describes the product of ethanol and acetic acid. Predictably, beer features a wide variety of ethyl esters—ethyl butyrate, ethyl hexanoate, ethyl lactate, and so on—as ethanol appears at significantly higher concentrations than any other alcohol found in beer. As a group, esters typically yield fruity aromas. Brewers can influence the specific mixture of esters found in a given beer through yeast strain selection and careful manipulation of fermentation parameters. Esters are often less volatile than other classes of flavor compounds, so techniques such as the covered sniff can help to draw them from the beer.

Ethyl Acetate

Threshold: 5–10 mg/L (ppm)

Common flavor descriptors: nail-polish remover, solvent

Ethyl acetate, formed from ethanol and acetic acid, is the most abundant ester found in beer by weight. Higher levels of this flavor compound often arise in beers that have elevated levels of either of its chemical inputs. High-alcohol beers often have increased ethyl acetate levels, as do beers fermented with *Acetobacter* or another acetic acid–producing bacteria. At low levels, ethyl acetate lends beer a light fruity aroma. However, at high levels, ethyl acetate takes on an unpleasant solventlike note (Fix 1999, 111).

Isoamyl Acetate

Threshold: 1.1 mg/L (ppm)

Common flavor descriptors: banana, pear, circus peanuts

Isoamyl acetate, formed from isoamyl alcohol and acetic acid, stands out as a key flavor in a number of different beers. Though present in most ales, this compound dominates the flavor profile of traditional German-style weissbier and often appears at high levels in beers fermented with Belgian yeast strains. One interesting note on isoamyl acetate—you will never encounter this flavor in beers fermented with *Brettanomyces*. While Brett can produce a wide variety of other esters, it will break down any isoamyl acetate it encounters (Yakobson 2012). In fact, some lambic producers use the disappearance of isoamyl acetate as an indicator of Brett activity in their maturing stock.[1]

Phenols

Phenols describe a wide group of different flavor compounds from a variety of sources. Yeast strains exhibiting the POF phenotype give us clove-like phenols such as 4-vinylguaiacol (4VG), while *Brettanomyces* and other wild yeast species can also

---

[1] Frank Boon, telephone conversation with author, June 5, 2020.

produce barnyard-like 4-ethylphenol (4EP) and 4-ethylguaiacol (4EG). However, several non-biological sources can yield phenolic compounds as well. Wood smoke, a complex mixture of various combustion products, contains several odor-active phenols like guaiacol and syringol. Vanilla flavor, whether from vanilla beans or oak barrels, primarily arises from vanillin, a phenolic aldehyde. And chlorophenols, chlorinated phenolic compounds commonly resulting from the presence of chlorine or chloramine in brewing liquor, impart powerfully unpleasant antiseptic or medicinal aromas.

4-Vinylguaiacol (4VG)

Threshold: 300 µg/L (ppb)
Common flavor descriptors: clove

4-Vinylguaiacol (4VG) provides the clove flavor found in traditional German weissbier (isoamyl acetate provides the banana flavor). 4-Vinylguaiacol forms when yeast breaks down ferulic acid, a compound that occurs naturally in the cellular walls of barley and wheat. Some breweries will use a ferulic acid rest during mashing to liberate additional ferulic acid, providing more starting material for yeast to convert to 4VG. Even without using this mash regimen, a traditionally produced wort will contain high enough levels of ferulic acid to yield significant levels of 4VG, assuming the brewer selects a POF$^+$ yeast strain for fermentation. If your brewery manages multiple strains of yeast and produces beers in these styles, you may want to train your panelists on this flavor. Identification of 4VG in a batch ostensibly fermented with a POF$^-$ yeast strain can offer an early warning sign of cross contamination of your yeast.

4-Ethylphenol (4EP)

Threshold: 300 µg/L (ppb)
Common flavor descriptors: barnyard, Band-Aid®, smoky

While *Brettanomyces* produces a wide variety of different flavor active compounds, the ethyl phenols, namely, 4-ethylphenol (4EP) and 4-ethylguaiacol (4EG), often receive most of the attention when discussing the typical Brett flavor profile. In the wine industry, winemakers specifically watch for evidence of 4EP as an indicator of *Brettanomyces* contamination. In addition to the genes associated with POF$^+$ yeast strains, *Brettanomyces* has a gene that encodes for the enzyme vinylphenol reductase, allowing Brett strains to convert 4-vinylphenol to 4EP and 4VG to 4EG (Tchobanov et al. 2008, 213). Although the flavor descriptors typically applied to 4EP do not necessarily sound pleasant, low to moderate levels of these flavors can offer a pleasant point of complexity in Brett-fermented beers. However, even if a brewer intends to feature Brett character in a beer, high levels of 4EP can become offensive. Regardless of whether your brewery handles *Brettanomyces*, training on 4EP can help your panel recognize biological contamination in a beer. Just as 4VG can indicate cross contamination of your yeast strains, 4EP can signal infection by *Brettanomyces* or other wild bacteria and yeast.

Tetrahydropyridine (THP)

2-ethyltetrahydropyridine (ETHP) and 2-acetyltetrahydropyridine (ATHP)

Threshold: ETHP – 150 µg/L (ppb); ATHP – 1.6 µg/L (ppb)
Common flavor descriptors: mousy, mouse urine, cereal, corn tortilla chip, fresh baked bread

The term tetrahydropyridine (THP) actually describes a group of compounds responsible for mousy, corn chip–like flavors in beer: 2-ethyltetrahydropyridine (ETHP), 2-acetyltetrahydropyridine (ATHP), and 2-acetylpyrroline (APY). Maillard reactions cause ATHP and APY to appear naturally in many grain-based baked goods, hence the flavor descriptors relating to those foodstuffs (Adams and De Kimpe 2006, 2299). To date, most studies of THP have focused on the presence of these compounds in wine, though brewers are paying increasing attention to these flavors as the popularity of mixed-culture beers has grown in recent years.

In beer, THP can be produced by *Brettanomyces* and certain lactic acid bacteria species. According to the currently proposed mechanism, *Brettanomyces* combines the amino acid lysine with either ethanol or acetaldehyde to produce ATHP. Brett will then slowly reduce ATHP to ETHP, effectively eliminating the flavor from the beer as ETHP is nearly 100 times less flavor active compared to ATHP (Grbin et al. 2007, 10877). Oxygen and metal ions such as iron also play a role in the production of ATHP (Snowdon et al. 2006, 6470). This may explain why fruited sour beers seem especially prone to THP development, as fruit additions often both introduce oxygen and raise the concentration of metal ions present in the beer.

Depending on the pH of the surrounding environment, ATHP can exist in two different forms. At standard beer pH, ATHP is not odor active; however, when we take a sip of beer containing ATHP, the higher-pH environment of our mouths converts the ATHP to its odor-active form, allowing us to perceive its flavor retronasally. If attempting to detect THP in a beer, instruct panelists to take a sip of the beer, move it around their mouths for 5–10 seconds, and then swallow, paying attention to any flavors that arise in the aftertaste. As with any flavor, some panelists may have trouble perceiving THP depending on their personal sensitivity to these compounds.

Brewers of mixed-culture beers and fruited sours often bottle condition their beers and will typically wait to release the beer until it is clear of any noticeable THP flavor (see p. 158). The biochemical conversion of ATHP to ETHP proceeds slowly, sometimes taking as long as six months, so most brewers will test their beers every two weeks or so throughout bottle conditioning until they find it absent of THP.

Metallic

Iron, copper, manganese, etc.
Threshold (as ferrous sulfate): 2.7 mg/L (ppm)
Common flavor descriptors: tinny, copper, blood

Metallic character in beer often results from the presence of metal ions such as iron, copper, or manganese. Metal ions can make their way into a beer through a variety of different channels, but the most common scenario involves brewing liquor contaminated with dissolved metal. If present in your source water, you can identify metallic flavors in the water itself. In many breweries, brewers begin each shift by tasting the brewing liquor to assure that it is free of any flaws, and some breweries even send water samples to their sensory panels for daily assessment. Other possible sources of metal ions include contaminated diatomaceous earth used to filter beer and rusted or corroded metal equipment within the brewery. Additionally, exposed brass in draught systems, which is common in systems using chrome-plated brass components, can impart metallic flavors to beer.

Oxidation Flavors
Oxidation covers a vast constellation of different chemical processes that occur in beer as it ages, and while some of these reactions involve interaction with molecular oxygen (O_2), a number of oxidation reactions do not actually require the presence of oxygen gas. Oxidation reactions inevitably occur over time in the closed system of a keg or can of beer.

With the wide range of reactions possible, there is still much we do not know about the many compounds that drive oxidation flavor in beer. As such, you should not attempt to reduce oxidation down to a couple of flavor compounds; frankly, it is not possible. Mimicking oxidized beer using added flavor attributes presents serious challenges. I shall discuss two of the more important, specific oxidation-related attributes, but to best train your panelists on oxidized beer you should use samples of aged beer, rather than trying to artificially recreate a facsimile of the effects of time.

trans-2-Nonenal (T2N)
Threshold: 50–250 ng/L (ppt)
Common flavor descriptors: paper, cardboard

For a number of decades, trans-2-nonenal (T2N) was presented as the oxidation compound in beer. In 1970, researchers identified T2N as responsible for the cardboard flavor present in oxidized beer, leading future studies of beer oxidation to focus almost exclusively on this compound (Vanderhaegen et al. 2006, 357). This mentality has shaped the way people discuss oxidation in beer today. Although the oxidation of beer can result in a host of different flavor changes over time, many tasters solely associate oxidized flavor with notes of paper and cardboard, common lexical markers for T2N. While T2N certainly can appear in old beer, the emphasis placed on this one compound has done a disservice to the broader industry's understanding of oxidation flavor.

trans-2-Nonenal forms during wort production and exists in finished beer bound to proteins. As beer ages the T2N–protein complexes disassociate, increasing the amount of T2N flavor detectable in the beer. Higher temperatures lead to faster rates of T2N disassociation, as well as the more rapid development of other oxidation flavors. Neither dissolved oxygen nor oxygen exposure of finished beer has an impact on T2N levels (Lermusieau et al. 1999, 29). Some sources claim that oxygen exposure of packaged beer (e.g., serving beer using an air compressor) will lead to paper and cardboard flavors in the beer.

This is simply not true. Exposure to oxygen will certainly wreak havoc on a beer, producing a host of negative oxidation flavors, but it will not lead to increased levels of the papery or cardboard-like notes associated with T2N.

Damascenone

Threshold: 25 µg/L (ppb)

Common flavor descriptors: berries, artificial grape, tobacco, black tea

Compared with *trans*-2-nonenal, damascenone appears in oxidized beer much more frequently. Damascenone forms in the natural world when plant material degrades, or, as Dr. Bill Simpson likes to say, "When green things go brown."[2] In beer, damascenone flavor shows up in oxidized hoppy beer. In highly hopped styles such as IPA, this flavor can develop within six to eight weeks, even faster if the beer is stored at elevated temperatures. After training on this flavor, most tasters realize that they have encountered it innumerable times before but have not been able to put a name to it. If your brewery produces hoppy beers, understanding this flavor can help your panelists recognize changes in your beer as it ages, and can serve as a useful touchstone when performing shelf-life assessments of your beers.

---

[2]   Bill Simpson, "5-Day Practical Beer Taster Training Course" (workshop), Chicago, IL, October 14, 2015.

6

QUALITY CONTROL TESTS

Every sensory test you run should start with a question. Perhaps you want to know whether changing the hops in your IPA improves the flavor profile, or how long your stout can remain in the trade before beginning to taste off. Before you set out to serve samples to panelists, you must first define the question that you hope to answer, as the nature of your question will dictate the test you use.

When starting out, some panel leaders attempt to build their testing protocols by consulting a list of the many different tests available to them. Triangle tests, duo-trio, paired comparison, descriptive analysis, tetrad, difference from control—the list goes on and on. "Naturally," they think, "if all of these tests exist, I should be using all of them with my panel, or at least most of them . . . right?" Overwhelmed by the sheer number of options, panel leaders can get bogged down, losing sight of the guiding principle that each test should be designed to answer a specific question. As a result, a panel leader may end up running unfocused or unnecessary tests that ultimately do not help the brewery make better decisions.

Existing sensory textbooks do not offer much in the way of guidance to the new panel leader. Most tests detailed in these texts fall into one of three categories: affective tests, which measure a taster's preference or overall enjoyment of a sample; descriptive tests, which describe the individual characteristics of a sample in great detail; and difference tests, which test whether two distinct samples exhibit sensory differences.

While each of these test types can play a role in a brewery sensory program, they do not answer questions relating to quality control.

On the level of the individual batch, our quality control–focused question might be, "Is this batch fit for release?" or, more specifically, "Does this batch fall within its brand specifications?" At first glance, difference tests (e.g., the triangle test) may seem like a perfect candidate for answering product release questions by allowing you to test each batch to see if it presents differences from the ideal brand profile. However, as we will explore in the next chapter, difference tests turn out to be too exacting for this task. Most beer brands on the market show subtle sensory differences from one batch to the next. Even large, highly automated breweries experience minor batch-to-batch variation. Difference tests know no nuance—while they can effectively tell you whether a sensory difference exists, they cannot tell you whether that difference actually matters. As your panelists grow familiar with each brand over time, they will likely be able to identify differences between two batches of a given brand, even if both batches are deemed perfectly fit to sell.

For the purposes of product release and routine quality control we need a different sort of test. Most sensory textbooks only devote a single chapter to the topic of sensory quality control, and the recommendations and test types mentioned usually reference a 1992 book called *Sensory Evaluation in Quality Control* (Muñoz, Civille, and Carr 1992). Muñoz et al. cover four different methods: difference from control, comprehensive descriptive testing, quality ratings, and the in/out method.

The first method, difference from control, works by comparing a series of production samples against a single control sample and scoring each sample based on the magnitude of the difference between it and the control, often using a scale of zero to ten (0 representing no difference and 10 representing extremely different). This method works well with products that can be produced to an exacting standard and have a long shelf life, such as cosmetic products (Lawless and Heymann 2010, 411). The flavor profile of beer changes rapidly over time, even under ideal storage conditions. Using the difference from control method requires you constantly replace your control and assumes that you can repeatedly produce an identical control sample, which falls outside the realm of possibility for most breweries. Thus, this method is impractical for use in beer quality control.

The second method, comprehensive descriptive testing, consists of scaling the levels of all key attributes of the product. This method provides a tremendous amount of data and requires a correspondingly tremendous time investment in panelist training, often in excess of a hundred hours. Additionally, this test works best for products that vary within a limited set of key parameters; beer makes a poor candidate for this type of testing due to the wide variety and large number of potential key attributes.

The third method, quality ratings, involves rating each sample on a quality scale established internally by the brewery. Quality scales differ from company to company—some companies use a nine- or ten-point scale while others opt for a five-point scale—but they all follow the same general principles. The different points on the scale represent differing levels of quality, often defined using qualitative descriptions. Typically, these scales will have a cutoff point, below which samples either require further examination or get outright rejected. However, defining the quality scale presents a significant difficulty. For example, what does a six mean on a nine-point quality scale? How should a panelist consistently distinguish between a three and a four? Although this method can work well if panelists understand the scale,

panelists often use the scale in different ways, leading to high scoring variability, particularly with more complex products like beer (Muñoz, Civille and Carr 1992, 108).

The fourth method, the in/out method, offers the simplest setup and execution. Panelists assess a group of samples against a set of product specifications and simply state whether the sample is in specification or out of specification. This method requires the least amount of training of any of the methods and also places less of a cognitive demand on panelists, allowing them to assess a greater number of samples within a single session. However, if used exactly in this format, you will not receive any indication from your panelists as to why they rejected a given sample.

In the beer industry today, most sensory experts advocate using a modified version of the in/out method known as true-to-target (TTT) or true-to-brand (TTB) testing. To my knowledge, this sort of test was popularized in the brewery setting by Lauren Limbach at New Belgium Brewing in the early 2000s.[1] In a true-to-target test, panelists evaluate each sensory modality (i.e., appearance, aroma, taste, and mouthfeel) against a target profile, marking each modality as either true-to-target ("TTT") or not true-to-target ("not TTT") before rendering an overall TTT or not TTT assessment for the sample. For any modalities marked not TTT, panelists should provide feedback on the characteristics that led to that decision. The TTT/not TTT responses of panelists provide you with a straightforward way to evaluate the performance of samples, while their feedback can offer direction if trying to troubleshoot an issue. The TTT test is elegant in its simplicity, quickly and efficiently allowing you to answer questions in a wide variety of different scenarios. This test method will form the foundation of your sensory program.

True-to-target testing sometimes goes by other names, true-to-brand testing being the obvious alternative, but also the in/out method referenced above, as well as pass/fail testing, stop/go testing, and go/no-go testing. However, I recommend sticking with true-to-target when explaining this test to panelists. True-to-brand can work as well, though I

---

[1] L. Limbach, telephone conversation with author, April 14, 2020.

prefer -target because this method can be used for far more than evaluating beer for product release, as we shall explore in chapters 11 and 12. In any case, inherent in most of the other names is rejection of the sample, which can carry other considerations. If panelists think that selecting "stop" or "no-go" represents a vote against selling the beer, they may be hesitant to give that response. When introducing this test, you can explain to panelists that marking a beer as not TTT will not necessarily prevent it from going out to the trade, but instead will flag it for further evaluation. While the linguistic difference is subtle, the psychological effect can be profound. Since someone higher up the chain will ultimately make the decision on whether a beer should be held back, panelists should be insulated from this type of language to allow for unencumbered and unbiased assessment of each beer.

THE TRUE-TO-TARGET TEST

As its name implies, the true-to-target (TTT) test compares a sample against a pre-specified target. Your first applications of this test will assist in product release, in which case the targets will be brand profiles for each of your beers. Your panelists will generate these targets themselves using the description test (see p. 87), which will ensure they are intimately familiar with the brand profiles. As your program grows, you will find many applications for the TTT test, from questions related to product shelf life to recipe changes and ingredient evaluation. The specific targets that you use will change depending on the question you are trying to answer, but the general format of the test will remain the same. (An example target can be found in figure 10.1 on p. 120.)

When presenting samples to panelists for sensory evaluation, you want to give them as little information as possible regarding the identity of the sample to avoid biasing their assessments. In most tests, samples should be presented with nothing more than a sample code. However, the nature of the TTT test requires that panelists know the identity of the sample. Beyond this though, any details specific to the sample, such as batch code or packaging type, should be kept from panelists. When panelists sit to taste, they should have access to the target descriptions, whether on printed sheets or through the interface that they use to enter their responses. Over

time, panelists will become quite familiar with target descriptions for brands and products that you test regularly, but you should still make the descriptions available in case a panelist wants to reference them.

When a panelist performs a TTT test, they should evaluate each modality individually against the relevant target description. I recommend having panelists evaluate in the following order: appearance, aroma, taste, mouthfeel. You can have your panelists work through the modalities in a different order if you prefer, just make sure that the order stays the same from test to test. The ballot that panelists work with will depend to some degree on the data collection tools that you have at your disposal (see Managing Data on p. 101). Ideally, you should ask panelists to indicate whether each individual modality is TTT or not TTT. If possible, panelists should also have the ability to give feedback on each modality. If a panelist marks a modality as TTT they do not need to provide feedback. However, if they mark a modality as not TTT you should require them to give a reason as to why they marked the sample that way. While the binary TTT/not TTT responses will help you flag problematic samples, feedback from your panelists can help determine what follow-up testing or troubleshooting steps should be taken.

In addition to requesting whether each individual modality is TTT, you should also ask panelists to assess whether the overall sample is TTT. Sometimes one of the modalities of a sample may be not TTT but the sample overall still features all of its key characteristics. In such a case, the sample should still be marked as TTT overall. For example, on a product release panel, a panelist could find that the body of a given batch of IPA was a little bit lower than expected and consequently mark the mouthfeel as not TTT. However, if the overall description of that brand defines the key characteristics as aromas of grapefruit and mango and high bitterness and the sample still meets that description, they should mark the sample as TTT overall, assuming the body is not way out of line. Some panelists may find this a bit confusing, so make sure to explain it carefully during training, and try to provide them with concrete examples if possible. Also, take care in crafting the "overall" section of each target profile, making sure to note all of the characteristics that must be

---

Product Release True-to-Target ballot for American IPA

Panelist Instructions: This is a product release sample of American IPA. Assess each modality against the target description. For each modality, please indicate whether the modality is true to target (TTT) or not true to target (Not TTT). For any modalities that you mark as not TTT, please describe how the sample differs from the target description. You may also provide feedback on modalities marked TTT if desired. After assessing each modality, make a final assessment of whether the sample overall is TTT or not TTT.

Target Appearance: Dark gold color, low haze, white foam

☐ TTT　　　　☐ Not TTT

Feedback:

Target Aroma: Grapefruit, mango, and orange, with slight green onion, pineapple, toast, and water cracker aromas

☐ TTT　　　　☐ Not TTT

Feedback:

Target Taste: High bitterness, low sweetness, no sourness

☐ TTT　　　　☐ Not TTT

Feedback:

Target Mouthfeel: Medium body, medium carbonation, low astringency, no alcohol warmth

☐ TTT　　　　☐ Not TTT

Feedback:

Overall Target: Must feature dark gold color, low haze, grapefruit aroma, mango aroma, orange aroma, and high bitterness.

☐ TTT　　　　☐ Not TTT

Feedback:

---

in place for the sample to pass. This section carries more weight than any of the others, as you will analyze your panelists' overall ratings of each sample to inform decisions.

ANALYZING TRUE-TO-TARGET DATA

You will quickly learn that each and every sample that passes through your panel will typically have a couple of panelists (or more, depending on the size of your panel) that mark the sample as not TTT, even for samples perfectly within specifications. Human panelists are variable instruments; their responses will vary from time to time depending on their personal sensitivities, their mood, what they ate for breakfast, and a host of other factors that influence their sensory experience. This is why you use a group of panelists rather than just one or two tasters. Over time, you will be able to calculate the average number of not TTT responses that your panel gives for each sample and this average will help establish a baseline for your panel. However, even with this average in place, how do we determine how many not TTT responses signals an issue? To answer this question, we turn to a common quality control tool: statistical process control (SPC) charting, or control charting.

Many industries use SPC charting to monitor the way a process behaves over time. In addition to analyzing sensory data, SPC charting can be used to monitor any quantitative aspect of the brewing process, from mashing efficiency to IBUs measured in a given brand (fig. 6.1). Control charts monitor a given variable and use the mean and standard deviation of that variable to set up control limits, which act as guardrails to alert you when an issue arises. (For more details on descriptive statistics like mean and standard deviation, see pp. 79–80.)

True-to-target tests rely on a specific type of SPC chart known as a *p-chart*, also known as a percent defective chart or sample fraction nonconforming chart because they are most commonly used to measure defective units produced by a manufacturing process (Montgomery 2009, 290). Because they accommodate binary pass/fail-type data, p-charts are well-suited for analyzing true-to-target tests. A p-chart tracks the percentage of not TTT responses that your panel gives each time they assess a sample. This percentage value is termed p, or the p-value. Consequently, the value of p for a given sample will vary between 0 (indicating no responses were not TTT) and 1 (indicating all

responses were not TTT). The p-charts track your panelist's responses over time—in order to establish the mean and standard deviation for your panel, you will need to plot at least 10 samples. When getting started with product release testing, you can begin by creating a p-chart tracking the performance of every sample tested regardless of brand, effectively creating a brewery-wide p-chart. As you accumulate more data, you may eventually switch over to using individual p-charts for each brand, though, for the sake of simplicity, it is fine if you decide to stick with using a brewery-wide p-chart to track all of your beers.

In addition to tracking the value of p from batch to batch, a p-chart will typically contain two lines that help to quickly evaluate data at a glance. The first line is the mean or *center line*, which is the average p-value across all samples assessed. This corresponds to the baseline number of not TTT responses that your panel gives on average. This mean value is then used to calculate the position of the other key line, which is the upper control limit (UCL). The UCL defines the percentage of not TTT responses that indicates a problematic sample—any p-values plotted above the UCL indicate the sample warrants further evaluation. The UCL sits above the mean line at a distance calculated based on several parameters from your data set. For more information on how to calculate the UCL, see sidebar "Building Your Own p-Charts."

Most control charts feature a corresponding lower

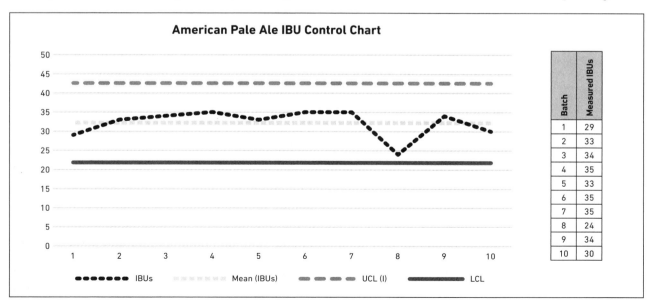

Figure 6.1. Example control chart mapping measured IBUs across 10 different batches of a brewery's American pale ale. Based on the values of this data set, the mean (center line) falls at 32 IBUs, with the lower control limit (LCL) and upper control limit (UCL) calculated as 22 and 43 IBUs, respectively.

BUILDING YOUR OWN p-CHARTS

Although they may look complex at first glance, p-charts are relatively easy to put together. If you use any sort of sensory software to collect and manage your data, the program will likely include tools to generate these sorts of charts automatically. However, if you want to create your own charts manually, you can do so using Microsoft Excel® or similar spreadsheet software.

The only data that you need to collect during a panel session is the number of "TTT" responses and the number of "not TTT" responses (see table below). From those two values, you can calculate all of the other values necessary to construct your p-chart. First, calculate the total number of responses (equivalent to the number of panelists present) by adding the number of TTT responses to the number of not TTT responses. You can then calculate the p-value for that sample by dividing the number of not TTT responses by the total number of responses.

To create the final chart, you will also need to calculate a few values for the dataset as a whole. To calculate the value of the center line (CL), find the weighted average (denoted as $\overline{p}$) of the p-values across all of the batches within the data set by dividing the total number of not TTT responses (130) by the total number of panelists across all sessions (488). In the case of the example data from figure 6.2, $\overline{p}$ is 0.27. Calculating the upper control limit (UCL) requires a couple of steps but is still relatively straightforward. First, find the standard deviation (σ) of the p-values within the data set by using the formula

$$\sigma = \sqrt{\frac{\overline{p}\,(1 - \overline{p})}{n_i}}$$

where σ is the standard deviation, $\overline{p}$ is the weighted average of p-values, and n_i represents the number of panelists present at a given session. (Note that since the number of panelists can vary from session to session, both the standard deviation and the UCL shown on your p-chart may vary slightly from session to session.)

Then, use the CL value and the standard deviation to find the UCL using the following formula—multiply σ by three and add that to $\overline{p}$ (0.27) to find the UCL at each point on the chart:

$$UCL = \overline{p} + 3\sigma$$

With the CL and UCL calculated, you can plot your p-values, allowing you to rapidly assess the performance of a given sample (see figure 6.2). And while this may seem like a lot of work, you should only need to set up these calculations once. If using a program like Excel, you can design the calculations to incorporate new data as they are added, which will automatically adjust both the CL and UCL as you perform additional tests. As your sensory program matures, you may wish to periodically refresh your calculations of the CL and UCL by excluding old data, perhaps data older than six months. Over time, your panelists' abilities will grow and the composition of your panel may change, so it is best to base the calculations of your control lines on recent data that accurately represent your current panel. *Continued on page 75.*

control limit (LCL) line, set at the same distance away from the center line as the UCL, but below it rather than above it. With a p-chart that measured defective units, you might assume that a low level of defects would be a good thing. However, assuming that no process changes have been made that would lead to an improvement, values far below the average typically indicate an issue in sampling processes or training of assessors. In the case of a sensory panel, a streak of extremely low p-values would likely represent a situation in which the sample targets were too permissive or the panelists had become complacent and were simply passing every sample. Most sensory p-charts will not feature an LCL, largely because the calculated value for the LCL would likely be negative (depending on your weighted average and standard deviation). However, if you begin to notice a

Continued from page 74.

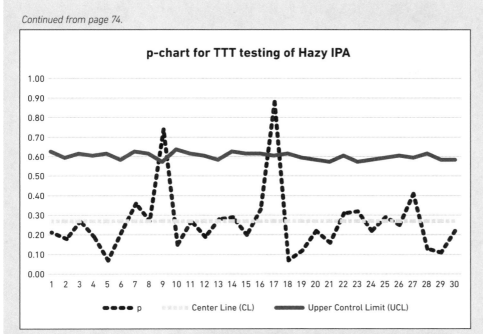

p-chart for TTT testing of Hazy IPA

●●●● p   Center Line (CL)   Upper Control Limit (UCL)

Batch	TTT responses	not TTT responses	Total # responses	p-value (not TTT responses/ total responses)
1	11	3	14	0.21
2	14	3	17	0.18
3	11	4	15	0.27
4	13	3	16	0.19
5	14	1	15	0.07
6	14	4	18	0.22
7	9	5	14	0.36
8	11	4	15	0.27
9	5	14	19	0.74
10	11	2	13	0.15
11	11	4	15	0.27
12	13	3	16	0.19
13	13	5	18	0.28
14	10	4	14	0.29
15	12	3	15	0.20
16	10	5	15	0.33
17	2	14	16	0.88
18	14	1	15	0.07
19	15	2	17	0.12
20	14	4	18	0.22
21	16	3	19	0.16
22	11	5	16	0.31
23	13	6	19	0.32
24	14	4	18	0.22
25	12	5	17	0.29
26	12	4	16	0.25
27	10	7	17	0.41
28	13	2	15	0.13
29	16	2	18	0.11
30	14	4	18	0.22
SUM	358	130	488	-

Figure 6.2. Example of a p-chart for a brewery's Hazy IPA, showing p-values obtained across 30 batches of the brand. Based on the values of this data set, the center line (CL) lies at p-bar(USE SYMBOL)=0.27. The upper control limit (UCL) varies slightly from batch to batch based on the total number of panelists that assessed each sample, but hovers around 0.6. The p-values for batches 9 and 17 clearly exceed the UCL, and so these batches would be subjected to further evaluation.

number of very low *p*-values across a short period of time (indicating that nearly all panelists are consistently passing every sample), you should slip a few spiked samples into your TTT test flights to make sure that panelists are paying attention.

ANALYZING DATA WITH A SMALL PANEL

If your panel consists of a small number of tasters, you may not find much value in using p-charts to analyze your data. With fewer than eight or so panelists, each taster's response carries too much weight for this format to function well. Just one or two panelists flagging a sample as not TTT can produce wild swings in your calculated *p*-values. With small panel sizes, you may want to approach TTT test analysis from another perspective. A small panel can produce results that, while perhaps not statistically significant, still allow the brewery to make informed decisions. If you have a panel of five

tasters and four out of five panelists mark a sample as not TTT, that is a pretty strong indication that that sample warrants further investigation. With a small panel, you can develop your own guardrails in terms of what percentage of panelists flagging a beer sounds the alarm, but you may also find value in the following hybrid method.

Instead of processing responses without the panelists' input, have the panelists taste samples individually and then follow up with a group discussion of the results. This way, panelists do not influence each other's perceptions during their initial tastings, but the group can build a consensus opinion together by comparing the results of different panelists. If you elect to go this route, you should carefully moderate the panel discussion to ensure that particularly assertive panelists do not strong-arm the more soft-spoken members of the group.

7
AFFECTIVE, DESCRIPTIVE, AND DIFFERENCE TESTS

While quality control should be the primary focus of your new sensory program and the true-to-target (TTT) test will be your main tool, other types of sensory tests can help you answer additional questions about your beers. What style of beer should we make next? Which hops should we use in our new IPA? Which of these two prototype saisons should we release? You can answer these questions using affective tests, which assess the hedonic qualities of a sample to determining how much tasters enjoy a given beer. What does this beer taste like? How can we describe and encapsulate this brand's flavor profile? These questions are best addressed by descriptive tests, which allow you to develop robust targets for use in TTT testing. Can we change the base malt of our porter without producing a noticeable flavor difference? Will dry hopping at a higher rate change the aroma of our New England IPA? These sorts of question can be answered through difference testing, though that is not the only possible approach.

Each time you set out to answer a question, try to think like a scientist setting up an experiment. Scientists employ the scientific method in an attempt to better understand the world around us. Regardless of the discipline, whether chemistry, psychology, physics, or physiology, the tools remain the same. Hypothesis formulation lies at the core of experiment design, followed by hypothesis testing using observation and measurement. Ultimately, that is what you are doing—using sensory science to try to better understand your brewery's beers. In order to analyze the results of sensory experiments, we often turn to statistics to help us process and understand our data. Before exploring other types of sensory tests not covered in chapter 6, let us briefly visit the world of statistics to see how statistical principles can affect the way we design our tests.

BASIC STATISTICS AND EXPERIMENTAL DESIGN

Statistics tend to get a bad rap, with many people considering statistics simultaneously boring and confusing. This reputation is understandable, if not entirely deserved. Between the complex formulas and the endless charts of values related to different statistical tests, statistics might seem like something best left to academia. But when properly used, statistics can help you make better decisions. Statistics may sound scary, but I promise that we are only going to cover the most essential topics and you will make it out on the other side unscathed. And while a small brewery may not have enough panelists to achieve statistically significant outcomes in their tests, the core concepts discussed here still underpin the setup of any successful sensory experiment. Understanding these topics will help ensure that you design your sensory tests properly, even if you do not actually perform much statistical analysis on your results.

Now you might be wondering how we use statistics and what we use statistics for. One of the most basic applications of statistics is simply to summarize and report data. If you calculate the mean, or average, of preference scores your panelists gave for a certain beer, you are producing a summary statistic—a single number that communicates information about a test you ran. We use various summary statistics of a data set, such as the mean, median, and standard deviation, to communicate different types of information.

While most people know how to calculate the mean, standard deviation warrants a bit more explanation. I am not going to go into how to compute standard deviation by hand, primarily because it involves a rather tedious series of individual calculations and you can skip all of that by using a spreadsheet program to perform the calculations for you. Within any given data set, the standard deviation will tell you how widely distributed your panelists' responses are. Take for example the data for two samples evaluated in figure 7.1. The two samples yield the same mean, but their standard deviations are very different. In sample 1, panelist ratings fell across a wide range of different values, whereas in sample 2 the panelist ratings clustered around the mean, with no responses other than 5, 6, or 7. Simply put, a smaller standard deviation represents a more closely aligned set of ratings from your panel of tasters.

Panelist	Preference Score	
	Sample 1	Sample 2
1	7	5
2	5	5
3	6	6
4	6	6
5	4	6
6	8	5
7	7	6
8	2	7
9	5	5
10	7	6

	Sample 1	Sample 2
Mean	5.7	5.7
Standard deviation	1.77	0.67

Figure 7.1. Example data sets illustrating how two samples may return the same average score (mean) but very different standard deviations. In this case, the ratings for sample 2 are much more closely aligned, as shown by the lower standard deviation.

Beyond helping us summarize and report data, we also use statistics to analyze our data. We often employ statistical analysis to help rule out chance variation. When analyzing the results of a difference test, we use statistical analysis to demonstrate that any differences observed by our panelists resulted from actual sensory differences. When we deem a result "statistically significant," we are not making a value judgment of the importance or relevance of this result. Rather, this phrase simply indicates a very low likelihood of obtaining such a result due to chance (see sidebar "Statistical Significance versus Practical Significance").

In the same vein, we often lean on statistical analysis to compare different values to determine whether they are far enough apart to indicate a real difference. This may seem trivial, but remember that sensory testing uses human subjects as instruments and humans are inherently variable. If we take the pH reading of two different samples using a pH probe and find that one has a pH of 4.2 and the other a pH of 4.4, we can state with confidence that the first sample has a lower pH than the second. If we ask our panel to rate the acidity of two samples on a scale of one to five and the first returns an average value of 4.2 while the second returns a value of 4.4, we cannot confidently state that one is more acidic than the other without further analysis. Because humans are less accurate and more variable than a mechanical instrument like a pH probe, we use statistical analysis to determine whether small differences are actually significant or just represent the natural variability of the panelists.

Ultimately, statistics are a risk management tool. In addition to summarizing the results of our tests, statistics tell us how confident we should be in those results. Depending on the gravity of a decision, you may need to adjust the way you execute your tests and analyze the results to ensure you answer the right question.

Decision Errors

Unfortunately, statistical analysis of sensory data cannot prove that your conclusions are correct. Statistical calculations rely on probability analysis, effectively predicting the odds that you would encounter your results by chance. For example, if you had 16 out of 18 panelists correctly identify the different sample on a triangle test, the calculations you perform

STATISTICAL SIGNIFICANCE VERSUS PRACTICAL SIGNIFICANCE

Whenever a discussion of statistics arises, mention of statistical significance is usually not too far behind. However, even people who routinely use statistical analysis sometimes lack a firm understanding of what statistical significance does and does not mean, and the term is often misused. In most experiments, researchers test whether variation in one variable has an impact on another. For example, a brewer may want to determine whether boiling their wort for an additional 15 minutes increases hop utilization and leads to higher IBUs. A statistically significant result would mean that, if the brewer did see an increase in the IBUs of the finished beer, it is highly likely that the increase resulted from the longer boil rather than some other factor or simply random variation in their measurements. In plain terms, statistical significance means that the results you obtained indicate a real relationship and are not just due to chance.

What statistical significance does not tell you is whether the results of an experiment actually matter. People often conflate statistically significant with the general definition of "significant," assuming that it means the results are important or meaningful. However, certain experimental parameters can lead to results that, while statistically significant, do not mean very much at all. For example, a difference test like the triangle test often requires large numbers of panelists to achieve statistically significant results. But as you increase the number of panelists, your test becomes more sensitive, increasing the probability of finding a difference even if that difference is vanishingly small. If you are trying to find a substitute for the base malt of your American pale ale, does it matter if a triangle test with 100 panelists finds a statistically significant difference between the two samples? Not necessarily, and I will explain why.

Statistical significance matters most in research settings, where scientists use tests to better understand the way the world works. In the brewery, we are more concerned with practical significance, sometimes defined as the magnitude of a change. To go back to the example above, whether panelists can detect a significant difference between the two batches made using different base malt is not necessarily the information you need. If you change one of the ingredients in a recipe, your primary concern should be whether or not the new beer still falls within the target brand profile and can pass a true-to-target test.

Some panel leaders at smaller breweries may get discouraged by their inability to produce statistically significant results, but do not let yourself fall into that trap. While you may not be able to write a research paper based on the results of your panel, that is likely not your goal. Small panels can still yield practically significant results that enable the brewery to make better decisions.

determine the probability that 16 out of 18 panelists would choose the correct sample if there was, in fact, no difference. And while the probability of such an extreme outcome in the absence of a real difference is extremely low, it never reaches zero. Since probability plays a role in these types of tests, the possibility always exists that your conclusions may be wrong, even if you run your experiment correctly and analyze your data by the book.

Using the difference test as a model (where two distinct samples are tested to see whether they exhibit sensory differences), we find four possible outcomes. But, as shown in figure 7.2, only two of these outcomes represent a correct assessment. We hope that our panelists as a group correctly identify a difference when a

		Reality	
		Products are different	Products are not different
Based on Sensory Data	Difference reported	Correct	Type I error (false positive)
	No difference reported	Type II error (false negative)	Correct

Figure 7.2. An illustration of how type I errors (false positives) and type II errors (false negatives) can arise in a difference test.

difference does exist between two samples. Likewise, we hope our panelists collectively do not identify a difference if no difference exists. However, while our samples could be identical from a sensory perspective, we might encounter a scenario in which enough panelists guess correctly on the difference test, causing our data to indicate that a sensory difference exists. In this case, we state that the samples are different even though they are not. This scenario is a **type I error**, also known as a false positive. Conversely, our samples may exhibit a real sensory difference that goes undetected by the majority of our panelists. In this case, we would fail to detect a difference when we should have. This failure to note a difference is a **type II error**, or false negative.

Consider a hypothetical scenario in which a brewery wants to change the base malt used in their American pale ale by swapping the 2-row malt it currently uses with a 2-row malt from a different supplier. Suppose the brewery ran a difference test on trial batches made using the different malts, but a type I error occurred. This would mean that the panel data led the brewery to conclude the beers exhibited different flavor characteristics when a real sensory difference did not actually exist. In this case, the brewery might decide not to move forward with the new malt, when in reality it could have used the new malt without causing any sensory changes in the brand. If we suppose a type II error occurred when the brewery ran its difference test, then the panel data would lead the brewery to conclude that no difference existed between the beers when, in fact, they did display sensory differences. In this case, the brewery might decide to start using the new base malt, only to find after releasing it to a broader audience that dissatisfied customers can taste the change.

When designing a sensory test, you should consider the relative consequences of committing a type I or a type II error. Difference testing uses specific parameters to adjust the relative risk of each error (see Alpha and Beta Risk, p. 91), but even if you will not be performing statistical analysis on your results, thinking about what could happen if your panel data led you astray can help you avoid those outcomes to some extent. In the scenario described above, the type II error, which caused the brewery to release a product that led to unhappy customers, represents the greater danger of the two error types. To guard against this result, the panel leader may want to adjust the parameters of the test to limit the chance of a type II error.

You can apply similar reasoning to the way you analyze data from a product release TTT test. Consider the two types of errors that can be made: panelists can hold back a sample that does not present any flaws (type I) or panelists can pass a sample that does not match the brand profile (type II). When a sample fails a TTT test, usually the first step is to run the sample through your panel at a future session to verify the result. Thus, the consequences of committing a type I error are relatively low. On the other hand, a type II error allows a nonconforming batch of beer to leave the brewery—this clearly represents the more significant error of the two. Occasionally serving spiked samples to verify that your panelists will actually stop a flawed beer from being released is one way to limit the potential for this sort of error.

Setting Action Standards

When designing your sensory tests, you should also define the action standards for each potential outcome. A set of action standards describes the actions that you will take depending on the conclusions drawn from the test, giving you a concrete next step regardless of the outcome. In the example in the previous section, the brewery might decide that if testing shows a difference between the two types of malt then the action standard is to continue making the beer using the same recipe; if testing shows no sensory differences between the two types of malt, the action standard is to switch to the new malt. Defining the action standards prior to performing the test should not take long, and it allows you to think rationally and unemotionally about your options, ultimately leading to better decision-making.

Determining the action standards and thinking about the different error types for a diacetyl force test offers an example of this type of analysis in practice. In a diacetyl force test, we are heating a sample of maturing beer and smelling for the presence of diacetyl to determine whether we can cold-crash the yeast and move the beer to a packaging tank. When performing this test, which type of error carries the greater consequence? If we commit a type II error we will conclude that the beer still contains α-acetolactate (the precursor to diacetyl) when it is actually clear of this compound. In this case, we can give the beer an extra day in the tank before testing it again. This costs us a bit in terms of efficiency but should not hurt the brewery too much. If we make a type I error, we will conclude that the beer is free of α-acetolactate when there is actually still some present. We subsequently cold-crash the beer and send it to the bright tank for packaging. Conversion of α-acetolactate occurs very slowly at low temperatures, so in this scenario the beer will likely make it into package without any perceivable diacetyl present. However, given a few weeks, all of the remaining α-acetolactate will be converted into diacetyl, resulting in a buttery, diacetyl-laden beer out in the trade.

Even though most breweries never run any sort of statistical analysis on the responses to a diacetyl force test, quick analysis of the action standards for each outcome shows that a type I error carries far greater consequences than a type II error. Thus, brewers are generally careful about concluding that a sample is free of α-acetolactate and will usually err on the side of caution by letting a borderline case spend an extra day in the tank rather than pushing a potentially problematic beer through. Using this sort of error analysis can help improve decision making, even when statistical analysis does not come into play.

We have now adequately covered the statistical knowledge necessary to design, run, and analyze most types of sensory tests. Properly setting up a difference tests requires a bit more statistical knowledge, but most small breweries will not find much use for difference testing for reasons that I shall cover in the section on difference testing (p. 88). However, if your program will be running difference tests, or if you are simply curious and cannot get enough statistics, that additional information is covered there as well.

AFFECTIVE TESTS

Sensory scientists use affective tests to gain a better understanding of which products consumers like the most. Affective tests employ simple questions with straightforward protocols to accommodate untrained tasters. However, the use of untrained tasters coupled with extreme variation in individual flavor preferences necessitates a large pool of tasters. Most texts recommend a minimum of 75 participants, though a larger population size will further improve the accuracy of your results (Lawless and Heymann 2010, 8). For most product categories, performing consumer preference tests is a costly endeavor, as consumers must be incentivized to come to a location and spend time participating in the test. In this regard, many

breweries already have a specialized tool at their disposal that can be harnessed to perform this type of test—a taproom!

Although a brewery taproom does not offer ideal conditions for sensory analysis, the setting does accurately simulate one of the environments in which people typically consume your beer. To perform one of the following tests with your customers, simply hook up a keg of a new product to an unlabeled tap and offer each customer a sample to rate when they come to the bar. If performing this type of test with your customers, I recommend at least giving your customers the general style of the beer prior to asking them to rate the sample; this will help focus their tasting and better approximates the experience they would have when ordering the beer for the first time. You can collect results using an app like Sample Ox or through a Google form with an associated short link or QR code, making it easy for patrons to enter their assessment directly from their phones.

Most sensory texts recommend against using trained panelists to perform affective testing (Meilgaard, Civille, and Carr 2016, 314–15). The rationale behind this suggestion is that trained panelists may not be representative of your consumers, so their preferences may not accurately correspond to those of your paying customers. However, many people work in the beer industry because of their love of beer, so the preferences of your trained panelists may lie closer to the preferences of your customers than is typical in other consumer product industries. And if you do not have the ability to collect consumer preference data on your beers, performing an affective test internally when prototyping a new beer will still help inform the decisions that you make.

You should not administer affective tests following a sensory test such as a descriptive test or a difference test (Lawless and Heymann 2010, 306). When performing a descriptive test or difference test, panelists tend to operate in an analytical frame of mind, breaking the beer down into its individual components. Affective tests (e.g., hedonic tests and preference tests) often involve a more holistic evaluation of the beer, and most panelists will struggle to switch into a holistic frame of mind after carefully analyzing a beer. If you want to have panelists perform both types of tests on a beer—for example, a description test and a hedonic test—the hedonic test should come first.

Acceptance Test (Hedonic Test)

While sometimes known as the hedonic test today, the original framing of the acceptance test asked assessors to scale the degree of acceptability of a given product, hence the name (Lawless and Heymann 2010, 325). To perform the acceptance test, present the assessor with a single sample and ask them to evaluate the sample before rating it on a hedonic scale (see p. 176 for a sample acceptance test ballot).

While you can perform an acceptance test using whatever type of scale you prefer, most sensory scientists default to the nine-point scale shown in figure 7.3. This scale was originally developed by the US Army in the 1940s to assess the acceptability of foodstuffs prepared for soldiers (ibid., 326). When analyzing results using this scale, assign the numbers 1 through 9 to the categories on the scale, with "Dislike Extremely" corresponding to 1 and "Like Extremely" corresponding to 9. You can then average the numerical scores of your assessors to achieve an average hedonic score for the sample.

Category	Number on hedonic scale
Like Extremely	9
Like Very Much	8
Like Moderately	7
Like Slightly	6
Neither Like Nor Dislike	5
Dislike Slightly	4
Dislike Moderately	3
Dislike Very Much	2
Dislike Extremely	1

Figure 7.3. Acceptance test nine-point hedonic scale

Beyond assessing consumer liking of a single sample, you can also use the mean (average) hedonic scores generated by the acceptance test to compare multiple samples. In this case, present the assessor with each sample individually, asking them to rate each sample before moving on to the next one. You can then compare the sample means, using a parametric statistical method such as the *t*-test, to determine if tasters showed a statistically significant preference for one sample over the other(s). Evaluating samples in this way gives you

STUDENT'S *t*-TEST

The Student's *t*-test—today a fundamental tool widely used in statistics and hypothesis testing—was developed in the early 1900s by William Sealy Gosset, a chemist working at the Guinness brewery in Dublin. The *t*-test can be used to predict how well the characteristics of a small sample group approximate the characteristics of the larger population that they were drawn from. Gosset used the *t*-test to assess large batches of raw materials (e.g., malt) by analyzing a small number of samples drawn from the batch and extrapolating those results to predict the quality of the overall batch. At the time, Guinness barred its employees from publishing their research or discoveries to avoid giving away trade secrets, so Gosset published the test under the pseudonym "Student," which remains coupled to the name of the *t*-test today (Dodge 2008, 234).

In addition to estimating the characteristics of a large population based on a small number of samples, we can use the *t*-test to compare the means (averages) of two separate data sets to determine whether they differ significantly. In the case of acceptance testing, we use the *t*-test to determine whether the mean hedonic scores of two different samples show that panelists preferred one sample over the other at a statistically significant level. As with other statistical analyses, computing a *t*-test by hand is a tedious affair, but programs such as Microsoft Excel (with its dedicated T.TEST function) can make short work of the calculation.

preference data in addition to an understanding of how much assessors enjoy each of the samples tasted.

Preference Test

In preference testing, assessors evaluate two different products and then select the one they prefer. Sensory scientists like this test because it mimics the way consumers behave when comparing two similar products (Lawless and Heymann 2010, 305). Preference

AFFECTIVE, DIFFERENCE, AND DESCRIPTIVE TESTS

testing works best when you are prototyping multiple versions of a product and want to find out which one your customers like the most.

To perform a preference test, the taster should be presented with both samples simultaneously. Aside from their numerical labels, the samples should have no identifying information. Ask the assessor to interact with both samples and then select the one that they prefer. I recommend requiring tasters to select a preferred sample, even if they feel ambivalent about which one they like better. Some variations of this test include a "no preference" option, but it greatly complicates analysis of the results.

To determine whether or not the tasters show a statistically significant preference for one sample over the other, first identify the sample that received the most votes. Using table 7.1, compare the total number of votes the preferred sample received to the number listed in the X column corresponding to the total number of tasters that participated (N). If the number of votes received by the preferred sample is greater than or equal to the number listed in the chart, then your group of tasters has demonstrated a statistically significant preference for that sample over the other. If the preferred sample fails to reach the number of votes indicated in the chart, you cannot state with confidence that the tasters showed a clear preference for one of the samples.

One downside of using preference tests is that the results of the test only tell you which sample your tasters preferred—the results do not actually tell you whether your tasters liked either of the beers. For example, it is possible that, of two beers, tasters moderately disliked one sample but absolutely hated the other. If you decide to produce the preferred beer, you will still end up producing a beer that, on average, your customers moderately dislike. To avoid this potential outcome, I recommend using the acceptance test when asking consumers to hedonically assess your beers.

Ranking Test

The ranking test asks panelists to rank a group of samples in order from lowest to highest across a single parameter. The ranking test can be used as an affective test or a descriptive test. When used as an affective test, it functions similarly to the preference test, just with three or more samples. For example, you might ask panelists to rank a group of four different beers based on their overall preference, assigning a rank of 1

Table 7.1 **Preference test analysis**

Minimum value (X) required to show a significant preference given total participants (N)		
N	X	%
20	15	75%
25	18	72%
30	21	70%
35	24	69%
40	27	68%
45	30	67%
50	33	66%
60	39	65%
70	44	63%
80	50	63%
90	55	61%
100	61	61%
125	74	59%
150	88	59%
175	101	58%
200	115	58%
300	168	56%
400	221	55%
500	273	55%
1000	532	53%

N, total number of tasters; X, number of tasters selecting a given option; %, percentage of tasters selecting a given option.

Notes: Values for X based on α = 0.05. To calculate X for any N not listed here, use the formula

$$X = 0.98\sqrt{N} + \frac{N}{2} + 0.5$$

to their favorite and 4 to their least favorite. When used as a descriptive test, it can be used to rank samples in order of increasing intensity of a specific trait, such as sour taste, citrus aroma, or capsaicin heat. I recommend limiting the number of samples served in a single ranking test to no more than five or six; as the number of samples to be ranked increases, the difficulty of the exercise goes up exponentially.

Whether used as an affective or a descriptive test, the analysis of ranking test results is the same. The method described here follows Kramer's rank-sum test; this is one of the most straightforward ways to analyze ranked data, though other methods exist. After panelists rank the samples, add their ranks together to yield a rank sum for each sample. Next, find the difference between each pair of rank sums by subtracting the smaller sum from the larger sum of each pair. Finally, reference the rank-sum table found in appendix A to find the difference necessary to indicate a statistically significant difference between two samples. Figure 7.4 illustrates a worked example of ten panelists ranking three samples based on hedonic preference. For ten panelists and three samples, the rank sums must be different by at least 11 to indicate a statistically significant preference (see table A.1, p. 164). In this case, panelists showed a statistically significant preference for sample 2 over both samples 1 and 3, whereas the difference between the rank sums for samples 1 and 3 is too small to indicate a preference for one over the other.

	Sample rank				Samples	Difference
Panelist	Sample 1	Sample 2	Sample 3		1 & 2	11
1	2	1	3		1 & 3	2
2	3	1	2		2 & 3	13
3	2	1	3			
4	1	2	3			
5	2	1	3			
6	3	1	2			
7	3	2	1			
8	3	1	2			
9	2	1	3			
10	2	1	3			
SUM	23	12	25			

Figure 7.4. Example ranking test with 10 panelists ranking three samples in order of preference. In this test, a rank of 1 indicates most preferred and a rank of 3 indicates least preferred. The differences for each pair of samples is calculated by subtracting the smaller rank sum from the larger (e.g., for "1 & 2", 12 is subtracted from 23 to yield a difference of 11).

DESCRIPTIVE TESTS

Descriptive tests yield a wealth of information about a given product by asking panelists to assess the distinct characteristics that present in each different sense modality. True descriptive analysis testing is the most robust type of sensory evaluation that we can perform on a beer. When sensory texts discuss descriptive analysis, they refer to a method in which panelists scale each primary attribute of the product in question. While this method only requires eight to twelve panelists, the panelists must be highly trained on a broad scope of

SINGLE ATTRIBUTE SCALING TEST

While true descriptive analysis may lie out of reach, you can still perform scaling of individual attributes with your panelists. For example, imagine you are developing a new German-style weissbier and want to target high levels of banana aroma. If your brewers produced multiple test batches, you can use your panel to determine which sample shows the highest level of banana character. One approach is to use a ranking test, asking your panelists to rank the samples in order of increasing intensity of banana aroma. Alternatively, you can ask your panelists to directly scale the level of banana aroma present in each sample, perhaps using a scale of one to five.

Since your panelists have not been trained on a specific scale for banana aroma (or whatever other characteristic you might choose to evaluate), your results will not be as accurate as they would be if produced by a highly trained panel performing descriptive analysis. However, by asking panelists to focus on scaling just a single attribute, the data you obtain will likely be good enough to determine which sample exhibits the highest level of banana aroma. To analyze data from an attribute scaling test, average the panelist scores for each of the samples, and then use a parametric method like the Student's *t*-test to determine whether the scores are significantly different from one another.

different attribute scales. For example, if you ask panelists to perform descriptive analysis of a New England IPA, you want them to scale all of the important flavors present in the beer, such as grainy flavor, bready flavor, mango flavor, pineapple flavor, orange juice flavor, onion flavor, and so on. But in order to accurately scale each individual attribute, you must first train panelists to not only recognize each attribute, but also to accurately report different levels of each attribute in a precise and repeatable way. When you map this across all of the different attributes in all of the different types of beer that you produce, it can quickly add up to hundreds of hours of training required. As such, these sorts of descriptive analysis methods lie out of reach of all but the very largest breweries.

While you may not be able to train panelists to the levels necessary to perform true descriptive analysis, the type of information offered by this category of test is too valuable to ignore. Instead, breweries can use the description test, which is a scaled-down version of the full-blown descriptive analysis method.

Description Test

Alongside TTT testing, the description test will serve as one of the most useful tools in your sensory arsenal. When first getting your sensory program off the ground, you will use this test to build descriptive profiles of your beers, profiles that will serve as your brand targets when you begin performing routine TTT testing. You will also find value in using the description test to evaluate new beers as part of the brand development process (see chap. 13).

The description test methodology—developed by Lindsay Barr of DraughtLab—retains many of the features of the descriptive analysis method. Panelists still assess each sensory modality, covering distinct aspects of appearance, aroma, taste, and mouthfeel. Panelists will even typically scale most of the simpler attributes in the realms of appearance, taste, and mouthfeel, though using ordinal scales (i.e., ratings of "low," "medium," and "high") rather than interval scales. The key difference between the two tests lies in the way panelists assess aroma. Rather than scaling each aroma or flavor present in the beer, panelists instead use a check-all-that-apply (CATA) format, listing the key aromas that they perceive in the beer by drawing from an established, standardized lexicon. Through analysis of the panel's data, you can build a consensus aroma profile because the primary aromas tend to be cited by a greater number of panelists (see fig. 7.6 on p. 88).

To perform the description test, present panelists with a single sample without any identifying information. Although panelists may have some idea of the identity of the beer, especially if it is an existing brand or the panelist works in the brewery, you should avoid giving any information about the beer and you should instruct the panelists to assess the beer as if they have never tasted it before. If presenting panelists with multiple samples, instruct them to completely evaluate each sample before moving on to the next sample.

Panelists should assess each sense modality of the beer individually. I recommend having panelists assess the modalities in the following order: appearance, aroma, taste, mouthfeel. However, it is fine if you want to use a different order—just make sure that you present the test to panelists in a consistent manner. Similarly, it is up to you to choose the specific attributes within each modality that you want panelists to assess. Below, I list the attributes that I suggest tracking. I have marked attributes that I consider essential in bold; the other attributes on the list may offer less value depending on the types of beers you produce. However, if this list contains any attributes that you do not find useful, remove it from your personal list (e.g., if you only produce low-alcohol beers, you may decide that tracking the level of alcohol warmth does not matter). If desired, you can also add to this list from any of the attributes mentioned in chapter 4 (e.g., if you produce a beer with chili peppers as an ingredient, you should add capsaicin heat to the list of mouthfeel traits that your panelists assess). The attributes I recommend tracking are:

Appearance: **color**, **clarity**, foam retention, foam color

Aroma: general reporting of key aromas

Taste: **sweetness**, **bitterness**, **acidity**

Mouthfeel: **body**, **carbonation level**, astringency, alcohol warmth

Of all the modalities assessed, you will likely derive the most value from your panelists' evaluation of aroma. In the aroma section, instruct panelists to report aromas they perceive using descriptors from a standardized lexicon such as the Beer Flavor Map. Panelists should usually target a total of four to six descriptors per beer. In a description test, panelists do not need to report the levels of each aroma they perceive.

Analyzing the results of appearance, taste, and mouthfeel data simply involves aggregating your panelists' responses for each characteristic and selecting the response that appears the most frequently (see fig. 7.5 for an example). To build a profile of a sample's aroma, tally up the number of times each descriptor is mentioned by your panelists. Typically, a consensus will emerge among panelist responses. Descriptors with the highest number of votes represent the primary aromas, descriptors with a moderate number of votes represent secondary or background notes, and aromas with few votes can usually be ignored (see fig. 7.6 for an example).

Panelist assessment of sweetness level for beer 324	
Panelist #	**Panelist Response**
1	Low
2	None
3	Low
4	Low
5	Medium
6	Low
7	Medium
8	Low
9	Low
10	Low
Most frequent response	**Low**

Figure 7.5. Example of a panelist sweetness assessment in a description test.

DIFFERENCE TESTS

Flip through any sensory textbook and you will find a nearly endless list of difference tests. While the sheer number of distinct test protocols may seem daunting at first, they all serve the same basic purpose—to determine whether or not a statistically significant sensory difference exists between two different samples. However, due to potential variations in brewhouse processes and fermentation outcomes, brewing often does not take well to difference testing. Two different brewers at your brewery producing wort following the same general process may end up with slightly different finished beers if they do not perform all the steps in exactly the same way. Dialing in fermentation conditions presents an even greater challenge. Can you ensure that, every time you pitch yeast, you are achieving a consistent pitching rate with a consistent level of yeast viability and vitality and a well-controlled level of wort aeration? Variation in any of these parameters can lead to perceivable differences in the flavor profile of the finished beer. Most breweries exhibit some batch-to-batch variation, usually enough that experienced panelists can tell one batch from the next. Given that baseline level of variation, attempting to use difference testing to assess a recipe change or a process adjustment will not give you valid

Composite description										Aroma Profile for New England IPA	
Prominent mango, pineapple, and orange, with secondary notes of green onion, white bread, and water cracker.										**Descriptors**	**Total Count**
Panelist 1	**Panelist 2**	**Panelist 3**	**Panelist 4**	**Panelist 5**	**Panelist 6**	**Panelist 7**	**Panelist 8**	**Panelist 9**		Mango	9
										Pineapple	7
										Orange	7
Mango	Orange	Peach	Mango	White bread	White bread	Mango	Grapefruit	Pineapple		Green onion	6
Pineapple	Mango	Orange	Green onion	Mango	Green onion	Peach	Pineapple	Mango		White bread	5
Green onion	White bread	Mango	Black pepper	Pineapple	Mango	Water cracker	Orange	White bread		Water cracker	4
White bread	Water cracker	Pineapple	Grapefruit	Orange	Orange	Pineapple	Mango	Green onion		Peach	2
Orange		Water cracker	Pineapple	Green onion	Boiled egg	Green onion	Water cracker			Grapefruit	2
						Orange	Nail polish remover			Black pepper	1
										Nail polish remover	1
										Boiled egg	1

Figure 7.6. Example of panelists' aroma assessment in a description test. By counting the number of times each aroma is mentioned by your panelists, you can produce a representative description of the beer's aroma profile.

results. If panelists encounter a difference, you cannot confidently state that it resulted from the change you made rather than routine batch-to-batch variation.

In all likelihood, you will not use difference testing as part of your sensory program, but that is not a bad thing. In truth, difference testing is best suited to research settings or manufacturing environments where subtle changes can make the difference between consumer acceptance and rejection. In most breweries, these sorts of tests will not give us answers to the questions we are trying to use them for. Assuming that a difference exists, a difference tests will not tell us anything about the magnitude or the nature of that difference. If you want to make a malt substitution, you do not truly need to know whether the change produces a statistically significant sensory difference—what you are really after is whether the substitution produces a large enough difference that your customers might notice. And to answer that question we have other tools at our disposal, as we shall explore in chapter 12.

This is not to say that difference testing has no place in brewing. Larger breweries often use difference tests to test new ingredient suppliers, process adjustments, different types of packaging material, and more. If you want to use difference testing at your brewery, you must first validate the assumption that your brewery can reliably produce consistent beer. If your panelists can tell the difference between two "identically" produced batches, then there is no point in using difference testing. To validate your brewery's production capabilities, set up a few rounds of difference tests using two batches of the same brand. Ideally, you should conduct this type of testing at least a few times using several different batches of beer. If your panelists consistently fail to find a difference between different batches, you can safely begin to use difference testing with your panel.

As with any kind of test, be deliberate in the way you design your difference tests. Never attempt to adjust more than one variable at a time within a given experiment. If you see a difference after altering multiple variables, you will not be able to identify which of the changes you made produced that difference. Also, do not perform a difference test on two samples that clearly differ from one another. Difference testing is designed to find minute differences between samples, not to confirm the obvious.

Types of Difference Tests

While there are subtle distinctions between different types of difference tests, they all accomplish the same basic goal. Consequently, most sensory scientists use whichever difference test they prefer, oftentimes just the one they are most familiar with. Historically, the triangle test has been the most widely studied and applied difference test, and it remains the most popular test in use today. That said, some tests are more powerful than others, essentially allowing you to achieve higher accuracy with fewer panelists. The tetrad test in particular has begun to receive more attention in the past several years due to its greater efficiency and accuracy compared with other difference tests (Sanderson 2017, 184). Below, I cover the procedures for the two of the most common types of difference tests: the triangle test and the tetrad test. If you decide to perform any difference testing, you can select the most appropriate or appealing test from these two.

Triangle Test

To administer the triangle test, present panelists with three samples simultaneously. The samples should represent two different products: two samples of one product and one of the other. All three samples should be labeled with random three-digit numbers and presented with no other identifying information. Ask panelists to first evaluate each sample, and then to select the sample that differs from the other two. The triangle test features six possible sample orderings (AAB, ABA, BAA, BBA, BAB, ABB). Sample presentation should be balanced across panelists to limit bias.

Tetrad Test

To administer the tetrad test, present panelists with four samples simultaneously. The samples should represent two different products: two samples of one product and two of the other. All four samples should be labeled with random three-digit numbers and presented with no other identifying information. Ask panelists to first evaluate all four samples and then to pair the similar samples with one another. Like the triangle test, the tetrad test features six possible sample orderings (AABB, ABAB, ABBA, BBAA, BABA, BAAB). Sample presentation should be balanced across panelists to limit bias.

Some long-time users of the triangle test have begun to switch over to the tetrad test in recent years, as some

BALANCED SAMPLE PRESENTATION

Balanced presentation involves serving each possible combination of samples to an equal number of panelists (or as close to equal as you can get). For example, if you used 24 panelists to perform a triangle test, each different ordering of samples would be seen by four panelists. If you only had 22 panelists for the same task, four of the orderings would be seen by four panelists while two would be seen by three panelists. You should also randomize which panelists receive which ordering, and panelists should be aware that they could receive any possible presentation order, as this too will help reduce bias.

studies have indicated that the tetrad test can produce more accurate results in certain circumstances (Adjei 2017, 98). In many cases, the tetrad test can achieve similar levels of accuracy with fewer panelists when compared to the triangle test, which makes it more attractive to sensory scientists. One study cited a triangle test with 220 assessors having equivalent power to a tetrad test with only 65 assessors (Sanderson 2017, 184). Unfortunately, this increased discriminatory power seemingly only occurs when using a large pool of assessors, with the effect becoming dramatically reduced when the number of tasters falls below 30.[1] If you have a panel of 10–20 tasters, for example, you will not be able to use the tetrad test to achieve results similar to performing a triangle test with 30–60 panelists.

STATISTICAL ANALYSIS OF DIFFERENCE TESTS

Once you have learned how to use a statistical reference chart, evaluating the results of a difference test becomes a basic exercise, often taking no more than a minute to complete. And while statistical reference tables can look rather intimidating to the uninitiated, they are astonishingly easy to use once you understand what the various variables signify. However, in order to understand how to use these tables, we are going

to have to take a slightly deeper dive into statistical concepts. If you will not be performing difference tests and are not interested in this material, feel free to jump ahead to the beginning of chapter 8—you will not miss anything by skipping this section. Once again, I will try to keep the stats jargon to a minimum and we will make it through this together.

Hypothesis Testing

We often turn to statistical analysis to test the hypotheses of our sensory experiments. In fact, whether you know it or not, any time you set up a difference test you are performing a hypothesis test with the goal of determining whether a real difference exists. Formally, when setting up a hypothesis test you must define a null hypothesis and an alternative hypothesis. Rather than pursuing abstract definitions, let us look at an example to illustrate these terms.

Suppose Brewery A currently brews a cherry stout using Farmer Bill's Standard Cherry Puree. Recently, Farmer Bill himself approached the brewmaster at Brewery A to let them know about a new product: Farmer Bill's Premium Cherry Puree. Although the premium product comes at a premium price, Farmer Bill assures the brewmaster that as soon as they taste the new puree they will want to upgrade. Not one to just take Farmer Bill's word for it, the brewmaster decides to brew a batch of cherry stout using the premium puree to compare against their current stout brewed using the standard puree. Once brewed, the beers are delivered to Brewery A's sensory panel for difference testing.

In determining our two hypotheses, the null hypothesis (H_0) represents the status quo, that is, *we assume that the null is true unless we discover evidence to the contrary*. You should structure your alternative hypothesis (H_a) so that the null hypothesis and the alternative hypothesis cannot both be true. In the case of our two cherry stouts, H_0 states that there is no perceivable difference between the two samples, while H_a states that there *is* a perceivable difference between the two. Although taking the time to define your hypotheses before setting up a test may seem tedious, this step can help you catch errors in experimental design, ultimately saving time and money by preventing you from having to redo a test.

1   Cammy Beyts, Richard Haydock, and Neil Desforges, "Is the Tetrad Test always appropriate? Comparison of the Tetrad and Triangle methods using small panel sizes" (poster presentation, Waltham Centre for Pet Nutrition, Waltham-on-the-Wolds, UK, May 31, 2017), https://www.slideshare.net/WalthamCPN/is-the-tetrad-test-always-appropriate-comparison-of-the-tetrad-and-triangle-methods-using-small-panel-sizes.

The exact test Brewery A chooses to employ will affect the number of correct panelist responses needed to establish significance, but it will not affect the way that the brewery interprets the results. If enough panelists correctly distinguish between the two samples, then the brewmaster can reject H_0 and accept H_a, concluding that the two different cherry purees did produce a sensory difference in the finished beers. However, if the number of correct responses fails to reach this threshold, this does not automatically provide enough evidence to conclude that the two samples are not perceivably different. In statistical speak, we call this scenario "failure to reject the null hypothesis," a somewhat confusing double negative that warrants some unpacking. This statement implies that, while we failed to show that the samples were different, there does still remain the possibility that they exhibit sensory differences and that our panelists missed this difference. Proving that two samples are the same from a sensory perspective requires that we reduce the risk of committing a type II error (false negative), and often involves a significantly greater number of panelists.

Alpha and Beta Risk

Depending on the way we design our experiments, we can actually control the probability of making a type I or type II error. This control comes from adjusting two specific factors: alpha (α) and beta (β).

The probability of making a type I error (false positive) is referred to as alpha risk (α-risk). In most cases, the value of α will be set to either 0.05 or 0.01, which corresponds to an α-risk of 5% or 1%, respectively. Sometimes you will hear results of a test reported at a given "confidence level"—this confidence level is simply one minus α. For example, a confidence level of 95% corresponds to an α-risk of 5%. You can think of α-risk as the probability of making a type I error and the confidence level as the probability of avoiding a type I error. The confidence level denotes the probability that we do not declare a difference when no difference is present.

The α value used in a given experiment is not something inherent to the test, but rather something chosen by the experimenter based on the consequences of committing a type I error. When comparing two similar tests that were run with different α values, the key difference is that tests with a lower α will require a greater proportion of panelists to correctly note the difference between the two samples in order to report that a significant difference exists. For example, if you ran a triangle test with 25 panelists and wanted to report results at a 95% confidence level ($\alpha = 0.05$, α-risk = 5%), then 13 of the 25 panelists would have to successfully identify the different sample in order to report a difference. If you ran the same test with 25 panelists but wanted to report results at a 99% confidence level ($\alpha = 0.01$, α-risk = 1%), 15 of the panelists would need to correctly identify the sample. The value of α (and the corresponding confidence level) should always be chosen prior to conducting a test based on the scenario and the confidence level required, rather than choosing values that fit your data. In the previous scenario, if you had determined that the scenario required an α of 0.01 but only had 14 panelists correctly identify a difference, it would be inappropriate to go back and state that your results were significant at a 95% confidence level. Instead, you would stick with your α of 0.01 and report that the evidence was insufficient to reject the null hypothesis.

The probability of making a type II error (false negative) is referred to as beta risk (β-risk) and often takes on values of 20%, 10%, or 5% depending on the consequences of making a type II error. Analogous to the concept of α and confidence, the value of one minus β is called the "power" of the test. For example, a β-risk of 20% would correspond to a power of 80%. Power can be thought of as the probability of avoiding a type II error. Power represents the probability of correctly rejecting H_0 when a difference exists. Framed another way, power is the likelihood that a test for similarity will pick up on a difference if one is present. In tests attempting to show that two samples are different, 20% is considered an acceptable level of risk for a type II error, given that your assumption is that the two samples are different. However, in tests that attempt to show that two samples are the same, we decrease β-risk so that we do not erroneously declare two samples the same if they are actually different. Reducing β-risk typically requires greatly increasing the number of panelists used, which is one of the reasons why tests for similarity tend to be more cumbersome than tests for differences.

Balancing Alpha and Beta

Although the exact relationship between α and β is a bit of a mathematical nightmare, we can show without too much trouble that as the chance of a type I error gets smaller (α decreases), the chance of a type II error gets larger (β increases). If we select a small α, we are effectively stating that the difference must be clear to a large number of people in order for us to confidently state that a difference exists. If this is the case, we increase the chance of missing a real difference that is not recognized by quite as many panelists, corresponding to a large β.

LOW α AND LOW β

While it is possible to achieve very low values for both α and β, this generally requires re-cruiting ridiculously large numbers of panel-ists (sometimes several hundred). Fortunately, this scenario is often not something we want to pursue. With high numbers of panelists the sensitivity of your test increases. This may sound like a desirable outcome, but increased sensitivity just means that your large pool of tasters will be more likely to identify a sta-tistically significant difference between two different samples. Because your large pan-el is so sensitive, you increase the probability of finding a difference that, while statistically significant, is entirely trivial from a practical perspective. By using an adequately sized pan-el (between 30 and 150 assessors, depending on the specific test and the resources avail-able), you can achieve useful results without the headache of trying to find several hundred tasters to assess your beer.

A sensory scientist often must decide what values to set for α and β depending on the parameters of the experiment and, more importantly, the action standards that will be pursued based on the test's results. When determining whether to set a low α or a low β for a given test, you should lay out scenarios for what would occur if you made a type I or a type II error. We often use statistics to manage risk, so your α and β values should be set based on which type of error would be more costly to make. If a type I error carries a particularly bad result, α should be lowered. Similarly, if a type II error carries a partic-ularly bad result, β should be lowered.

Let us revisit our hypothetical cherry stout one last time to examine whether a type I or a type II error would pose more of a risk to Brewery A. With a type I error, the panel data indicates that a differ-ence exists between the beer brewed with standard puree and the beer brewed with premium puree. In this case, the brewery may decide to use the more expensive premium puree even though it does not produce a sensory difference in the finished beer. With a type II error, the panel data indicates that no difference exists when there is actually a real sen-sory difference between the two beers. In this case, the brewery misses the opportunity to potentially improve on the cherry flavor in its cherry stout. However, assuming that customers already like the beer, this may not represent much of a loss at all. While customers are quick to notice a drop in quality, they will not miss a potential improvement that never occurred. In this case, the type I error represents a significantly greater danger in terms of wasted resources; therefore, the brewery may decide to perform the difference test with a lower α, perhaps α = 0.01.

For an example of a case where a type II error would carry greater consequences than a type I error, refer back to the example of a brewery attempting to substitute the base malt of their American pale ale (p. 82). Perhaps the brewer wants to switch to a less expensive malt and hopes to confirm that the beer brewed with the cheaper base malt tastes the same as the current beer. Here, a type II error would mean that the panelists found no significant difference between the samples when a difference did exist, potentially leading the brewery to make the switch and send an inferior beer to market. In this case, the brewery may decide to reduce β to 0.10 or 0.05 to try to prevent this outcome.

In the realm of statistical hypothesis testing, most tests are designed with an α of 0.05 and a β of 0.20. These two standard values balance the risks of making either type of error. However, as demonstrated in the scenarios above, the values chosen for a particular hypothesis test should be chosen carefully based on the consequences of making a type I or type II error.

Similarity Testing: How Large of a Difference is Acceptable?

In consulting statistical reference tables for difference tests, you will encounter a couple of additional variables depending on the test you choose: P_d for the triangle test and delta (δ) for the tetrad test. Although these two variables convey distinct concepts, they both generally denote how large or obvious of a difference we are willing to accept between two samples in a similarity test. These variables acknowledge the fact that samples prepared in different ways will always exhibit some amount of a sensory difference and some tasters will be able to accurately distinguish between the two. In setting up a test for similarity, we first have to decide how large of a perceivable difference we are willing to accept while still concluding that the samples are "the same."

In triangle testing, P_d stands for the proportion of "true discriminators" in the population and typically ranges from 10% to 50%, with higher percentages representing a larger proportion of tasters that can consistently identify a difference between the two samples. In tetrad tests, δ represents an outright estimate of the magnitude of the difference between the two samples, with $\delta < 0.5$ corresponding to a small difference, δ between 0.5 and 1.0 corresponding to a moderate difference, and $\delta > 1.0$ corresponding to a large difference (Sanderson 2017, 187). In both cases, as we lower the value of P_d or δ (indicating that only a small number of people will be able to tell the two samples apart), the number of panelists required to demonstrate similarity increases significantly. For example, running a triangle test for similarity with an α of 0.10, β of 0.10, and a P_d of 50% requires a minimum of 15 panelists. However, if we can only tolerate a very small difference between the two samples, we might keep our α and β values fixed, but lower P_d to 10%—such a test requires a minimum of 348 panelists.

In practice, setting an appropriate P_d or δ for a similarity test requires a judgment call on how large of a difference you are willing to accept. However, very few breweries actually use difference tests to try to establish similarity, as the TTT test can accomplish the same goal with fewer panelists. If you decide that you would like to use difference tests to test for similarity, consult a resource such as *Discrimination Testing in Sensory Science: A Practical Handbook* (Rogers 2017) for the appropriate reference tables to analyze your results.

Using Statistical Reference Tables

Phew! If you have made it this far, you now possess the knowledge to appropriately analyze the data from a difference test. Equipped with this understanding, you will actually find the analysis relatively straightforward. If you had to manually perform statistical analysis on data from a discrimination test, you would be in for quite a tedious exercise. The variables and equations used change depending on what type of test you are performing, how many tasters you have, and what you set α and β to. However, most sensory scientists do not actually run involved calculations to determine statistical significance every time they run a test. Conveniently, these calculations have already been performed for a wide variety of different scenarios to cover different test types, different numbers of panelists, and different α and β values, and the results have been organized into statistical reference tables. Once you have set α and β, and P_d or δ, you can look up the minimum number of assessors needed for your test. And once you run the test, you simply use the values you already selected coupled with the number of tasters that sat for the panel to look up the number of correct responses necessary to indicate a statistically significant result.

For example, suppose you perform a triangle test with 25 panelists looking for a significant difference with α set to 0.05, then the calculated value of correct answers needed to establish significance is 13. If 13 or more panelists correctly identify the different sample, then you can say that a statistically significant difference

Table 7.2 **Minimum number of correct responses necessary to establish statistical significance in a triangle test (α = 0.05)**

Minimum value (X) required to indicate a significant difference	
N	X
20	11
25	13
30	15
35	17
40	19
45	21
50	23

N, total number of tasters; X, number of tasters correctly identifying the different sample exists between the two samples at a confidence level of 95%. If fewer than 13 panelists correctly identify the different sample, then your data fails to reject the null hypothesis. Analysis of these tests requires no calculations, only enough understanding of the underlying parameters to look up the appropriate value in a reference table (see table 7.2 for an example). Assuming that you can accurately predict how many panelists will show up, you can even pull the number of correct responses required in advance so that you can draw a conclusion as soon as the results of the test come in.

To use statistical reference tables, begin by identifying the table that corresponds to the test you want to perform. To determine how many assessors you need for your test, use α and β, and also P_d or δ depending on whether you are performing a triangle test or a tetrad test. To analyze a test designed to find a difference, use α coupled with the number of tasters used to determine the minimum number of correct responses that indicates a statistically significant difference. Statistical reference tables for the triangle test and the tetrad test can be found in appendix B.

8
PLANNING, EQUIPMENT, AND BEST PRACTICES

Prior to conducting sensory experiments with your panel, you should establish a set of standard operating procedures, ranging from guidelines for your testing space to detailed specifications for how you prepare samples. Close attention to the details of how you perform your experiments matters just as much as the training you provide to your panelists. By minimizing variability in the way that you conduct panel sessions, you can eliminate potential sources of variation and bias.

CREATING A SPACE

In small breweries, resources are often in short supply, and none more so than space. When discussing spatial requirements, most sensory texts detail the specifications for designing a permanent sensory space, complete with individual booths that have pass-through doors leading to a sample preparation area. If you have the means to set up such a space, that is great! Your program will certainly benefit from having a space within the brewery explicitly devoted to sensory work. However, a sensory program can still produce valid, useful results without a dedicated space or custom-designed booths. In fact, you probably already have most of the tools you need to begin running panel sessions. Do not let lack of space or fancy equipment prevent you from starting

a sensory program. Good sensory programs derive their strength from the quality of their protocols and panelists, not from the space in which sessions are held.

Location

At a bare minimum, your sensory program will require a testing room where panelists will evaluate samples and a preparation room ("prep room") where you can prepare samples out of sight of your panelists. Neither of these rooms need to be spaces dedicated solely to sensory work. For example, a brewery taproom can double as a testing room when not open to the public. Regardless of whether these spaces are used for multiple purposes or exclusively for sensory work, you should ideally locate the prep room close to the testing room to minimize sample transport.

If possible, place the testing room in a central location within the brewery to make it easy for panelists to get to the space. While it may seem trivial, reducing the distance that panelists have to travel to get to panel sessions will improve your attendance rates. That said, your primary concern in selecting an appropriate location is to find a testing space free of any potential sensory distractions. This means locating the testing space away from sources of noise or aroma. Try to find a space away from loud areas like the packaging line, loading docks, or common social gathering places.

Similarly, you probably want to avoid the areas directly surrounding the brewhouse due to the potential of competing aromas.

When selecting or designing the location of your testing and prep rooms, you should also take panelist traffic patterns into consideration. Ensure that panelists do not have to walk through the prep room on the way to the test room. Allowing panelists to see samples being prepared can bias their responses; for example, if they see spike capsules out on the counter or in the garbage, they will be on the lookout for flaws in the beers that you serve that day. As a general rule, you should not allow panelists to enter the preparation area around the time of a panel session, so make sure that the layout of your spaces does not require them to go through the prep room for any reason.

Resources permitting, you may want to establish a permanent, dedicated space for your sensory program. In such an installation, the prep room customarily sits adjacent to the test room. Inside the test room, booths line the shared wall between the test room and prep room, and each booth has some sort of pass through allowing you to transfer samples directly from the prep room to your panelists. If designing a space at your brewery, *Sensory Evaluation Techniques* by Meilgaard, Civille, and Carr (2016) is a good starting point as it provides details on typical specifications for sensory booths.

As your sensory program develops, you can consider adding additional spaces. Common additions include a room with a conference table to use for training sessions, meetings, or sensory sessions that involve round-table discussions. This room may also double as a waiting area for panelists prior to sitting for a panel. As your pool of panelists grows over time, you will invariably have more panelists than booth spaces. Having a comfortable space for panelists to congregate while they wait for a booth helps make the panel experience more pleasant for panelists. However, if this space is directly adjacent to the testing room, encourage waiting panelists to refrain from conversation because this can disrupt the panelists currently tasting.

Lastly, if you grow to the point where you have full-time sensory staff, dedicated office space within the sensory facilities will certainly makes their lives easier. The entire footprint of the space does not have to be that large, but investing in your sensory facilities will show that the brewery values the sensory work panelists do.

Testing Room Considerations

In the testing room, individual sensory booths are the gold standard (fig. 8.1). However, you do not need permanent booths to gather valid data. In early sensory labs, panelists generally performed their sensory assessments sitting around a conference table. However, studies showed that this configuration creates the potential for panelists to bias one another (Meilgaard, Civille, and Carr 2016, 30). In setting up your testing room, try to avoid placing panelists in an orientation where they can see one another during a testing session.

Figure 8.1. Sensory booths effectively isolate panelists from their surroundings, allowing them to focus entirely on the samples in front of them. Photo courtesy of Allagash Brewing Company.

The simplest solution is to position tables along the walls of the testing room so that panelists face the wall rather than one another. You can take it a step further by constructing temporary booths to better isolate panelists from one another and make it easier for them to concentrate while tasting. Rudimentary individual booths can be constructed from three pieces of plywood connected by hinges.

When setting up the testing room, make sure that you allow enough space for each panelist. Guidelines for constructing booths suggest a width of 27–32 inches (69–81 cm)—any less than this can make panelists feel claustrophobic (Meilgaard, Civille, and Carr 2016, 32). If working without booths, use this same guideline to establish the minimum spacing between panelists. If space is a limiting factor this will determine the number of booths or spaces you can accommodate and, hence, the number of panelists that can taste at one time. As your program grows, this can become a bottleneck, effectively constraining the number of panelists you can seat in a given session.

The testing room itself should be comfortably furnished to create a relaxing atmosphere. If possible, paint walls white or off-white to avoid distraction (Meilgaard, Civille, and Carr 2016, 38). Most sensory texts recommend lighting of 70–80 foot-candles (about 750–860 lux), equivalent to a well-lit office space (ibid., 35). You certainly do not need to go to the trouble of carefully measuring the level of illumination of your room, but make sure that each booth or tasting space offers even, shadow-free lighting. Climate control matters too, both for panelists' comfort and proper sensory operation. The temperature should be between 22°C and 24°C (72–75°F) and relative humidity between 45% and 55% (ibid., 38).

To maintain an odor-free environment, the testing room should be supplied with slight positive pressure from air passed through activated carbon filters. If possible, the carbon filters for the room should be easily accessible to facilitate regular replacement. Additionally, try to avoid the use of absorbent materials such as porous tile, carpet, drapes, or other fabrics, as these can absorb odors and re-emit them to the room.

Preparation Room Considerations

The prep room should be designed to facilitate easy and efficient preparation of samples for panel sessions. You can never have too much counter space or storage space, so make sure to factor sufficient amounts of each into the layout of the room. When planning out how much counter space you will need, ensure that you have enough space to accommodate both sample preparation and staging of sample trays.

You need both refrigerated and non-refrigerated storage within the prep room. You should store your beer under refrigeration, and you should set your refrigerator to run at serving temperature (around 5°C, or 40°F) so that you can serve beer directly from cold storage. You also need dry storage for flavor spikes, glassware, panelist snacks, and other supplies. If the brewery cannot allocate a room to serve as a dedicated sensory prep space, you can use large plastic tubs to store your panel supplies for easy transport.

Your prep space should include access to a sink and running water to facilitate easy cleanup following panel sessions. Additionally, if you use glasses to serve your samples you may want to install a dishwasher in your prep room; and, while it may seem obvious, do not forget to include garbage and recycling bins. Lastly, if you need to regularly serve your panelists samples of kegged beer, you may find it helpful to locate a direct draw kegerator in your prep room.

SUPPLY CONSIDERATIONS

Beyond creating a comfortable tasting space for panelists and an efficient preparation space, you will need some basic supplies to effectively run panel.

Sample Prep Supplies

For samples served from bottles and cans, you can pour directly from the package. You do not need to transfer the beer to another container prior to serving. Depending on your preference, draught samples can be served directly from the tap or can be transferred into a pitcher or other container prior to pouring into individual glasses. For fermentor samples, bright tank samples, or any spiked samples, you need to pour the beer into some sort of container before serving it to panelists. Most texts recommend using glass, stainless steel, or glazed ceramic vessels for mixing and storing samples because these will not transfer any flavor to the sample (Meilgaard, Civille, and Carr 2016, 39). Sealable glass or Pyrex media jars work best as they also help prevent loss of carbonation and volatile aromas. However, hard plastic pitchers work reasonably well if that is all you have. If you elect to use plastic pitchers, you should test them to ensure that they do not transfer any aromas to the beer, and you should limit residence time within the pitchers to help limit any possible flavor impacts.

Panel Administration Supplies

For ease of presentation and transport, use trays to serve samples to panelists. If you have booths set up with pass-through doors, make sure that your trays are small enough to easily slide through the doors from the prep room to the panelists. If serving panelists sitting at tables, the exact size of your trays matters less, although using trays will still help facilitate efficient service of samples.

When it comes to glassware, the ASBC states that you should use a straight-sided, cylindrical, 250 mL (8 fl. oz.) glass.[1] However, some panel leaders prefer to use glasses that taper inward, such as a wine glass

---

[1] *ASBC Methods of Analysis* (online), "Sensory Analysis 2. Test Room, Equipment, Conduct of Test (International Method)," approved 1986, rev. 2009 (St. Paul, MN: American Society of Brewing Chemists), p. 2, doi:10.1094/ASBCMOA-Sensory-2 (subscription required).

or tulip glass, or a snifter. Some panel leaders elect to use plastic glassware, but I recommend using glass as it prevents the possibility of flavor transfer and is more sustainable. Do not overthink this decision—choose a size and shape of glassware that you like and be done with it. Just make sure that your glasses can hold approximately three times the amount of sample that you will serve panelists, which allows panelists to swirl the beer without fear of spilling and also allows for aroma compounds to build in the headspace of the glass. Ideal sample sizes range from 50 to 100 mL (1.7–3.4 fl. oz.), so a glass that can hold around 250 mL (8 fl. oz.) works well in most situations. Beyond that, just be consistent—use the same type of glass any time you serve beer to panelists, including during training sessions.

Figure 8.2. At Allagash Brewing Company, samples are served in small snifter glasses with keg cap covers to limit distracting aromas when a sample is not in use. Photo courtesy of Allagash Brewing Company.

Many panel leaders serve their samples with a cover over each glass (fig. 8.2 shows an example). While you do not have to use them, sample covers can help panelists concentrate on each sample individually by blocking competing aromas from other samples. If you want to use sample covers, a few cheap and readily available options include petri dish lids or keg caps.

To label samples, you can either use place mats, wax pencils, or printed labels. Place mats placed directly on your sample trays allow you to supply key pieces of information about the sample, such as the brand identity for a true-to-target (TTT) test sample. Alternatively, depending on your data collection system, you may be able to provide all necessary information on your ballots (whether digital or paper),

and use place mats solely to number the samples so that panelists can keep them organized. If you prefer, you can use a wax pencil to write directly on the glass instead. Wax pencils are typically odorless and water-resistant, making them an excellent tool in this environment. You can also affix printed labels to your glasses to denote the identity of each sample, but removing printed labels from glassware can be a bit of a pain and over time can leave adhesive residue on the exterior of your glasses. Avoid labeling glasses directly with a Sharpie® or some other sort of marker, as the marker itself may leave behind distracting off-aromas.

SAMPLE LABELING

Most sensory texts specify that test samples should be labeled using random three-digit numbers to avoid biasing panelist responses. This is an important consideration in tests that involve comparing two or more samples, such as the preference test or any of the difference tests. Labeling samples for these sorts of tests using sequential numbers (1, 2, 3, . . .) or letters (A, B, C, . . .) implies a certain order, which can potentially distract panelists or distort their responses. In tests that involve panelists assessing individual samples, such as a true-to-target test, a hedonic test, or a description test, this consideration is less important. For these tests, you can still use random three-digit numbers to label samples, but you can also use sequential numbers if desired.

During a panel session, you should also provide panelists with palate cleansers, which can be placed directly on the serving tray. Always offer panelists an ample supply of water so that they can cleanse their palates in between samples. Unsalted saltines or table water crackers make for an effective palate cleanser as well (Lawless and Heymann 2010, 65).

Panel Booth Supplies

Although the testing area should be free from distractions, you might want to include some information for panelists to refer to during their sessions. Most importantly, panelists should have access to the brand profiles for any TTT tests that they perform. If you use a digital data entry platform, you may be able to supply this information within the data entry interface. If not, printed and laminated profiles for each brand should be available at each booth or tasting space. Beyond brand profiles, you may also want to include a flavor descriptor resource, such as the Beer Flavor Map, to assist panelists with their tastings.

MANAGING DATA

Returning to our best practices and principles for running a sensory program, one of the primary goals of any good sensory program is to gather actionable data. Regardless of the quality of your panelists and their training, or your own skill in designing and administering sessions, your program will only realize its full potential if you have a good way to manage the data it generates.

The first element of data management to consider is data collection, and the first decision you need to make is whether to collect data digitally or via paper and pencil ballots. Until recently, most sensory programs collected panelist responses using paper ballots. However, with the increasing prevalence of computers, digital entry has begun to become the norm. Digital entry offers a number of significant benefits over paper ballots. First, digital entry allows panel leaders to receive results from their panelists in real time, and, once properly configured, can yield rapid analysis of panel results, allowing decisions to be made more quickly. Digital forms eliminate the time spent transcribing paper ballots into a usable form for data analysis. Some sensory programs have reams of completed ballots that never get entered and analyzed due to the sheer amount of time required to manually process them. Furthermore, having panelists record their results directly into a digital form avoids the opportunity for mistakes, errors, and typographical errors during subsequent data entry.

In brewing and other industries, larger sensory programs often equip individual booths with computers to facilitate data entry into digital forms. However, to save money, you can exploit the fact that most of your panelists likely have smartphones, which they can use to interface with the digital entry platform of your choice. This does create the potential for panelists to be distracted during a panel session, but proper instruction on panel etiquette will usually curb

these sorts of issues. If you prefer that panelists not use their personal phones or you have a few panelists that do not have smartphones, you can keep a few tablet computers available for use in the testing room. This maintains flexibility in your setup of booths or tasting spaces, and it will certainly cost less than purchasing a computer for each booth or paying someone to do data entry of paper ballots.

A number of different companies offer sensory data management software, some of the best known being Compusense, RedJade, and DraughtLab. If you prefer the do-it-yourself approach, you can use Google Forms to create individual ballots for each panel you administer. While this route offers the lowest up-front costs and greater flexibility, the additional time spent designing ballots and manually cataloging and analyzing data can end up costing a brewery more than simply using sensory data management software. However, if budgetary constraints preclude the use of preexisting sensory software, something like Google Forms will certainly serve you better than paper ballots.

To use your own digital ballots, begin by creating template ballots for the different types of tests you will perform with your panel. (For sample ballots for different types of tests, see appendix C.) When administering a given panel, build a flight of evaluation forms from your collection of different template ballots, and finish by using a URL shortener or a QR code generator to produce an easy link for your panelists to access the ballots from their phone. Google Forms will allow you to collect panelist data in a spreadsheet, after which you will need to manually process the data for further analysis. Information on how to analyze each of the major types of tests can be found in chapters 6 and 7.

CONDUCTING A SESSION

The exact flow of a given session will depend largely on the specific tests or tasks pursued that day. However, the following general guidelines and best practices will serve you regardless of the exercises you perform with your group.

Panel Timing

When scheduling your panel sessions, try to set them up at a consistent time each day. Depending on the number of tests that you have to run within a given week, you may not need to run panel each day. In fact, when just getting started you may only need to run sessions once a week. Regardless of the frequency of your sessions, holding panel sessions at a set time will allow panelists to build it into their routine, making it easier for them to attend regularly. Most sources recommend scheduling sessions for late morning, prior to lunch. A start time between 10:00 and 11:00 a.m. works for most programs. Panelists should not eat, smoke, or consume strongly flavored beverages like coffee within 30 minutes of a panel session.[2] Planning your sessions within this time frame will allow panelists to comfortably enjoy breakfast or a morning cup of coffee without compromising their performance in the sensory booth. However, if another time of day works better for you and your panelists, go with that—the easier you make it for panelists to attend sessions, the better luck you will have maintaining panelist motivation. Whatever time you choose, just try to be consistent about it.

The number of booths or seats your tasting room has will limit the number of panelists that you can seat at one time, so you will often have to serve a given panel across a window of time. Most panel leaders will serve samples over a 30- to 60-minute period depending on the number of panelists they need to accommodate. Also, make sure that you start panel sessions on time, with samples ready to go as soon as the first panelists show up. Doing so shows your panelists that you respect their time. Once your panelists become familiar with the format of a typical panel session, their speed and efficiency will increase. Experienced panelists often take no more than five to ten minutes to complete a given panel.

Sample Preparation

Sample preparation begins with gathering your samples for the day. On the day of a given panel, you should begin the morning by collecting all samples of beer that your panelists will taste. For bottles and cans, pull packaged beer and relocate it to the refrigerator in your prep room. Move any kegs you plan to test to a draught system that will allow easy service, whether that is a kegerator in your prep room or another draught setup nearby. If your panel will taste any fermentor or bright tank samples, gather those in sealable glass or Pyrex media jars and place them in refrigerated storage in your prep room along with your other samples. If you do not have dedicated glass media jars for this purpose, growlers will work too.

When preparing samples, try to limit the amount of time that a container of beer spends open prior to being served to panelists. Your primary concerns here are loss of carbonation and dissipation of highly volatile aroma compounds such as certain sulfurs; many of the techniques we use in sample preparation are designed to minimize either of these outcomes. Keep cans and bottles refrigerated and sealed until immediately before you serve them to panelists. Similarly, keep sealed jars of fermentor or bright tank samples under refrigerated storage until immediately before you need to pour them. If using sealable media jars, you can safely prepare keg beer and spiked samples in advance of a panel session. However, if using pitchers, prepare these samples immediately prior to serving them. Work quickly and place prepared samples in the refrigerator as soon as they are ready. Keeping the samples cool will help minimize loss of carbonation.

To prepare samples of kegged beer, either pour individual samples directly from the tap or carefully pour the beer into a pitcher or media jar in a manner that minimizes foaming. Once you have poured your sample, seal the vessel if possible and place the prepared sample under refrigerated storage.

To spike samples, first determine the total volume of beer that you will be serving and set your desired spike rate. A rate of three times threshold is standard during training for most flavors but, depending on the test and the experience level of panelists, you may wish to use a level higher or lower than this. Most premade flavor standards are calibrated to produce a level of three times threshold when one spike capsule or vial is added to one liter of beer, so you may need to adjust the amount of spike material used depending on the volume that you plan to serve. Assuming that you do not need to hit an exact concentration (as will often be the case for this sort of use), you can just eyeball the amount of spike that you add if you need to use a partial spike. However, if you prefer to be more exact in dividing up a spike, you can create a stock spike solution by adding the spike to a set volume (e.g., 300 mL) of beer in a graduated container and then use a specific amount of that stock solution to prepare the spiked sample(s) for your panel.

To spike the beer, begin by pouring one-third to one-half of the total volume of beer into a pitcher or

---

2   *ASBC Methods of Analysis* (online), "Sensory Analysis 2," p. 2 (subscription required).

sealable jar. Pour gently along the side of the vessel to minimize foaming and loss of carbonation. Once the beer has settled and any foam has subsided, add the predetermined amount of spike material. Swirl the vessel gently to mix the spike into the beer and then add the remainder of the beer to the vessel, once again pouring gently to minimize foaming. Seal the vessel and swirl gently one last time before placing the sample back into refrigeration prior to use.

How Many Samples to Serve

Each time you set up a panel session, keep in mind the total number of samples that you will be presenting to panelists. Careful sensory evaluation requires great focus from your panelists and asking them to evaluate too many samples in one sitting can take a toll on both their accuracy and their psyche. If panelists consistently leave panel sessions feeling worn out from being forced to churn through too many samples, their motivation will suffer and they may stop coming altogether.

The number of samples you can reasonably serve in one session will depend on the cognitive demand of the tests performed as well as the overall intensity of the beers. The ASBC suggests that panelists can perform as many as 10 to 12 pass/fail or TTT tests in one sitting, but since your TTT tests will involve individual assessment of each sensory modality, I recommend serving a maximum of eight TTT samples within a single panel. As your panelists become more experienced you may be able to approach 10 or even 12 samples, but if you find yourself regularly serving this many samples to your panel you should consider whether you need to hold panel sessions more frequently or whether you might be oversampling.

As the tests become more intensive, the number of samples you can reasonably expect your panelists to evaluate goes down significantly. For tests like the triangle test, you should serve no more than six samples, that is, a total of two triangle tests. For description tests—some of the most demanding tests that you will ask your panelists to perform—serve no more than four samples within a single session. If putting on a panel with multiple types of exercises, keep these guidelines in mind and scale the number of samples accordingly. For example, if you had a few samples to evaluate for release using a TTT test and

a few prototypes that you wanted to evaluate using the description test, you could serve panelists four TTT samples and two description samples without inducing fatigue.[3] As breweries grow, the number of samples requiring evaluation invariably increases. Larger breweries will often segment panels by purpose, for example, running a panel specifically for product release samples and another panel for descriptive testing of new brands. As your program evolves, try to remain cognizant of the demands you place on your panelists and adapt the structure of your panel sessions to accommodate their needs.

Warm Up and Presentation of Samples

Before beginning a panel session, consider serving panelists a warm-up sample to help transition their brains into tasting mode and to limit potential first sample effects when they begin tasting the actual samples to be evaluated. Present the warm-up sample to panelists before they sit down to interact with the test samples. Good candidates for the warm-up beer include your flagship brand or one of your core brands that has recently passed a panel. Make sure that panelists know that the warm-up sample does not represent a gold standard and should not be used as a reference during their evaluation of other samples. You do not need to ask panelists to perform any sort of formal evaluation of the warm-up sample, just ask them to perform a quick TTT test in their head, briefly evaluating each modality. This will prepare the panelists for the rest of the panel. Ideally, you will collect the warm-up sample from panelists before they sit down so that they cannot use it while interacting with their other samples.

When preparing the samples for a given panel, begin by laying out the total number of trays you will need for that session and then lay out all of your glassware on those trays. Prior to the beginning of the panel session, you should only pour beer for as many trays as you have booth spaces. Then, as panelists arrive and begin tasting, pour a new tray for each tray that you send out. This ensures that the beer served to panelists only sits out for a short period of time prior to being tasted.

Temperature has a tremendous impact on a beer's flavor. Endeavor to serve your samples at a consistent temperature from sample to sample, session to

---

3   *ASBC Methods of Analysis* (online), "Sensory Analysis 2," pp. 2–3 (subscription required).

session, and week to week. The ASBC recommends serving beer at 12°C (54°F) for full flavor assessment, while using temperatures of 8°C (46°F) for assessment of mouthfeel.[4] However, in practice it is often more valuable to serve samples closer to the temperature at which your consumers will drink them. Of course, samples will naturally warm while panelists assess them, and if panelists wish to speed up this process they can cup the glass in their hands. To maintain a consistent serving temperature, run the refrigerators in your preparation area at 5°C (41°F) and leave all sample vessels under refrigeration until the time of service, pouring samples directly from the refrigerator.

Ensuring Panelist Safety

As panel leader, your number one responsibility is to ensure the safety of your panelists. Every time a panelist sits down to taste, they blindly imbibe an unidentified liquid, implicitly trusting that you will not serve them anything that can harm them. It is up to you to make sure that this trust is not misplaced. Safety considerations primarily fall into two categories: overall alcohol consumption and the condition of the samples themselves.

When serving panelists, whether in training or during a panel session, be cognizant of the amount of alcohol that you serve them, because you do not want to allow them to get intoxicated. If any panelists have concerns relating to alcohol consumption during the workday, remind them that you can provide a spit cup so that they can still effectively participate in the panel session.

Breweries are dangerous places, presenting nearly every possible type of hazard found in a manufacturing setting. Within the context of the sensory program, your primary safety concern lies with the condition of the samples you serve to panelists. Contamination can occur, particularly with cleaning chemicals like peracetic acid or caustic solution. Beer has an extraordinarily high buffering capacity; pH monitoring will not help you in this circumstance,

although caustic contamination may cause color or aroma changes. It is up to you as the panel leader to pay attention to these types of cues—you serve as the last line of defense between your panelists and a dangerous sample. Learn to recognize the signs of contamination, whether chemical or microbial. Take note if you have a beer that gushes when you open it or if a beer shows unexpected turbidity. If anything ever causes you to think that a sample might pose a risk to your panelists, do not serve it to them! It simply is not worth the risk. And if a consumer complaint ever comes in suggesting that the beer made them sick, do not blindly put it in front of your panelists. Find another way to analyze the beer.

Beyond making sure that your samples are fit for consumption, take a proactive approach to protecting your panelists. Part of the panelist onboarding process should entail soliciting information on any allergies to prevent serving a panelist something that might cause them harm. Be aware that preparing samples involves handling a lot of glassware that people will then bring into contact with their mouths. Avoid handling the upper third of glasses and use gloves while preparing and serving samples. Keep hand sanitizer on hand, both for yourself and for your panelists. Make sure that your glass cleaning procedures adequately sanitize glasses in between uses. And if you feel sick have someone else prepare samples. Do what you can to avoid spreading illness among your panelists.

Lastly, take steps to care for the mental and emotional well-being of your panelists. Hopefully, your brewery already fosters a company-wide environment of inclusivity and has zero tolerance for harassment. Regardless of existing policies, as panel leader you set the tone for what sort of behavior is and is not appropriate at panel sessions. Stand up for your panelists and listen to them when they come to you with concerns. Treat your panelists with care and respect and you will be rewarded with a committed team of panelists you can count on.

---

[4]   *ASBC Methods of Analysis* (online), "Sensory Analysis 2," p. 2 (subscription required).

9

SELECTING, TRAINING, AND MAINTAINING PANELISTS

Everything up to this point has been designed to give you the theoretical knowledge necessary to build and run a functioning sensory program. Now begins the exciting task of actually setting up your panel. We will start with the first step—actually selecting and training the people that will make up your panel. Initial panelist training will form the foundation of your panelists' sensory skills, setting them up for success during panel sessions. Do not be daunted by the need to train your panelists, and certainly do not let it prevent you from getting started. Some breweries assume that they have to invest months of training in their panelists to assure their mastery of a wide variety of different flavor attributes, and that panelists will not produce viable data until then. This could not be further from the truth! With a bit of initial guidance on tasting technique and brand profiles you will be ready to run product release panels. Your panelists' abilities will naturally improve over time through the experience they gain by regularly attending panel sessions. Do not forego the benefits of starting a sensory program because you do not believe you can train your panelists perfectly.

PANELIST SELECTION

Prior to training your panelists, you will first need to select them. In large organizations, the pool of potential sensory panelists sometimes far exceeds the needs of the company and their capacity for training, so prescreening tests are used to narrow the field. Most sensory texts extensively cover panelist prescreening, which often involves tests to identify the basic tastes and will sometimes feature other exercises, such as identifying key flavors relevant to the product category or performing a couple of triangle tests. While necessary in some situations, I recommend against setting up any sort of prescreening tests at your brewery.

For starters, prescreening tests may exclude panelists that have high potential but little experience. If I administer a chemistry exam to someone who has never studied it, they will undoubtedly perform poorly. However, this does not indicate that they are bad at chemistry or should never study the topic. It is simply a consequence of the fact that they have not been taught the content yet. Similarly, testing someone on their ability to distinguish among basic tastes when they have never been explicitly trained to recognize them will not tell you whether or not they have any aptitude for sensory work, especially given that most flavor evaluation revolves around identification of aromas.

From a more pragmatic angle, you probably do not need to prescreen your panelists because the pool of potential panelists is likely small enough that you will want to use as many of them as possible. Though assessments like the true-to-target (TTT) test will provide valuable information with just a handful of panelists, you should ideally use anywhere between 10 and 30 panelists. Typically, only large regional breweries can put up these sorts of numbers, and even some large breweries struggle to maintain that many active panelists. In your case, you want to train everyone willing to participate in sensory work with the hopes that they will make for effective panelists.

Understanding Panelist Differences

Although your panel will generate aggregate sensory data as a group, each individual panelist shapes the overall performance of the team. Each panelist will bring to the table different strengths and weaknesses in the form of differing thresholds to specific flavor compounds, but also in traits that extend beyond their sensory acuity. Panelist attitudes and personalities can contribute or detract from the overall success of your panel. Understanding each panelist as an individual can help you keep your panel functioning optimally.

With regards to sensory acuity, you will come to learn your panelists' individual sensitivities over time through both training and routine panel sessions. This

---

ASCENDING METHOD OF LIMITS TEST

Many sensory texts outline a specific test called the ascending method of limits test, used to determine a panelist's specific threshold for a given flavor compound. There are several different permutations of this test, the most common of which involves presenting panelists with a series of triangle tests where the compound in question increases in intensity as the test proceeds. Concentrations typically begin with some fraction of the reported threshold (e.g., one-quarter of the threshold value) and then increase in strength across each test by a factor of two. For example, the first triangle features two control samples and one sample spiked to one-quarter of the threshold value; the second has two controls and one sample spiked to one-half of the threshold value; the third has two controls and one sample spiked to the threshold value; and so on. Panelists typically perform six sequential triangle tests, with their threshold determined based upon the highest concentration sample they fail to identify. In some cases, additional triangle tests can be performed to either confirm or further narrow the potential threshold range of the panelist.

Most panelists are inherently curious to know how their sensory abilities stack up, so this test certainly appeals to the ego. Unfortunately, the value of the results obtained from this method rarely warrant the time invested to run the test. Panelist sensitivity can vary from day to day, so the threshold calculated on one day may differ from the threshold found on the next. Additionally, the threshold for a given compound varies depending on the background matrix, that is, the style or brand of beer. If you produce a Kölsch and a barleywine, your panelists will demonstrate remarkably different thresholds for any given compound across the two beers. This raises the question of what beer to use when performing the ascending method of limits test—or should you test a panelist's threshold for diacetyl in every single brand that you produce? Regardless of your decision, the results only report a panelist's threshold for one single flavor compound. Given all of the important flavor compounds in beer, you can easily find yourself devoting all of your sensory panel's time to determining each panelist's thresholds for a wide variety of compounds across a large number of beers. And the whole time you must remember that a panelist's thresholds can change from one day to the next!

The ascending method of limits test certainly has its uses within sensory science. Notably, the test offers one of the best ways to determine the average threshold value for a given compound by sampling across a large group of individuals. However, if used to gauge panelist performance or acuity, the test reinforces the misguided notion that attribute recognition matters more than any other sensory skill set. Instead, you want panelists to focus on developing familiarity with your core brands alongside a descriptive vocabulary of positive flavor traits, as these skills will serve them far better in quality control exercises.

knowledge will serve you well. If you know which of your panelists are most sensitive to acetaldehyde, when you receive complaints of acetaldehyde in a given batch you can run the sample through your panel and note how those sensitive panelists assess the beer to validate whether or not acetaldehyde is actually present. Similarly, if six out of eight of your panelists flag a sample for diacetyl but you know that the two outlier panelists have low sensitivity to that compound, you can be reasonably certain that the sample contains diacetyl.

A panelist's behavior can also affect their ability to adequately perform as a member of the group. Your panelists should be open to adjusting their behavior in the 30–60 minutes leading up to panel so that they can perform as well as possible. While a panelist who regularly smokes tobacco will likely not quit solely to improve their ability on a sensory panel, they should not smoke in the hour before a panel session begins. The same goes for coffee drinkers and drinking coffee. Similarly, you should encourage panelists to refrain from eating in the hour leading up to a session, as consumption of food will alter the way they perceive their samples.

Know too that your panelists' performance will vary from day to day based on their mental state, their diet, the weather, and a variety of other factors outside of your control. While you cannot account for each and every variable, aim to make panel sessions a comfortable and enjoyable experience, and understand that your panelists will have off days from time to time. During training, let your panelists know that their performance may vary from day to day, and that this sort of individual variation is completely normal. Equipping them with this knowledge can help prevent panelists from being too hard on themselves if they occasionally miss a sample.

What if You Have a Low-Acuity Panelist?

Although you should train your group before making decisions regarding a specific panelist's ability, you may have to contend with a panelist whose performance detracts from the overall function of your panel. Although some people consider themselves "bad tasters," rarely will you encounter a panelist who is universally insensitive to the aromas found in beer. Given proper training and opportunities to practice, most people find that their ability to perceive

specific flavors improves greatly. Nonetheless, you may encounter a panelist who has great difficulty using descriptive flavor vocabulary or perhaps consistently misidentifies certain flavor compounds, even after completing training.

In these cases, you should first provide some gentle coaching. Ideally, you will periodically meet with each panelist on an individual basis to discuss their performance, affirm their participation, and solicit input on ways you can improve their panel experience. With a low-performing panelist, you can use such a meeting to identify weaknesses and suggest ways to improve their skills. In many cases, if someone is already investing their time and energy to regularly attend panel sessions, they are likely motivated enough to put in a bit of additional work to improve.

Unfortunately, in some cases, you may find a panelist whose performance fails to improve and falls short of your standards. Even in this case, I do not necessarily recommend outright removing them from the panel. Rather, your choice on how to proceed should be informed by their overall attitude regarding panel participation. If you have a low-performing panelist who enthusiastically shows up to each session, speaks highly of the sensory program to employees outside of the panel, and helps to maintain a positive atmosphere among panelists, you potentially lose a lot by excluding this person from panel participation. In some cases, panelists view their time spent in a panel session as a brief respite from the stresses of their other work-related tasks, and the boost in attitude they receive from participating in panels results in overall improved performance in other areas of their job. In such a case, you do not have to remove a low-performing panelist, you can simply exclude their responses when processing the data from a given panel session. Avoiding the potential drop in morale is often worth the minor effort required to prepare one more set of samples.

In my opinion, a panelist should only be removed from the group if their attitude or behavior negatively affects the performance of the panel as a whole. If a panelist consistently skips panel sessions without any repercussions, other panelists may begin to wonder if their own participation is truly necessary. If a panelist consistently distracts other panelists by making noise or talking during a panel session, you may need to remove this panelist to preserve a positive environment for the rest of your group. In these instances, you should meet with the offending panelist before outright removing them—in some cases they may be going through a tough period of time or may not have realized how their behavior affected others around them. However, if a misbehaving panelist refuses to reform their behavior you should remove them from panel for the sake of the other panelists.

How Many Panelists Do You Need?

For small breweries, typical recommendations for panelist numbers may seem entirely unattainable. A minimum of 10 panelists to plot p-charts for a TTT test? Or 18 to even run a triangle test? For some breweries, this might exceed its total number of employees! However, while higher numbers of panelists can increase the statistical significance of your results, you can still produce useful sensory data with a small pool of panelists. If you have a group of five panelists and four panelists fail a sample in a TTT test, you do not have to plot the results on a chart to know that you should take a closer look at that batch.

But how small is too small? A panel can still generate useful information with as few as two panelists, as independent, blind tasting of samples still offers a quality check on your beer. Regardless of the size of your brewery, each batch of beer should be tasted by a group of people before going out into the market.

PANELIST TRAINING

Designing your panelist training program offers you the exciting opportunity to shape the knowledge base of your panelists. Think of building your training program as a sort of choose-your-own-adventure activity. Yes, there is an abundance of topics you *could* cover with your panelists, but try to stick to those areas that offer the greatest value. Focus on the essentials to get your panel up and running; you can always offer ongoing training to improve your panelists' skills over time. Many panelists find training one of the most rewarding portions of panel participation, relishing the chance to grow their sensory skills and knowledge. And the time you do spend on training will pay significant dividends over the lifetime of your program in the form of stronger, more reliable results.

The following sections cover a wide variety of topics you may want to consider presenting to your panelists during training. The first two—general knowledge and skills, and brand training—are the most essential.

If you have time, you can cover some of the other areas as well, but those first two topics will equip your panelists with enough knowledge to get your panel off the ground. Over time, you will learn what sort of information and training best serves your panelists and, if desired, you can develop a more robust training program for onboarding future panelists as your sensory program grows.

General Knowledge and Skills

When you convene your first training session, begin by explaining to your future panelists why sensory data matters in the first place. This serves several purposes. First, it helps to build a culture of sensory within your brewery, reinforcing the idea that sensory plays an integral role in maintaining the quality of your beers. Second, it helps reaffirm the importance of sensory-related activities to any managers or department heads who allow their employees to take time away from core duties to participate in sensory work and training. Lastly, and perhaps most importantly, it reminds the trainees themselves that what they are doing matters. In turn, this will increase their focus in both training and tasting sessions and will help sustain panelist motivation in the long run.

Specify what the sensory program does (or what you hope it will do, if you are just getting started) and explain how data that the sensory panel produces informs decisions made around the brewery. Most brewery employees take pride in the beers made by their company—allowing them to see that their participation in a sensory panel has a direct bearing on whether beer leaves the brewery can serve as an incredibly powerful motivator.

In your first meeting with new panelists you should also lay out general ground rules for both sensory training and sensory panels. Ask trainees to refrain from wearing heavily scented products on days that they will be participating in sensory training or evaluation, and cover expectations relating to panelist behavior before and during panel sessions. Most importantly, emphasize that panelists should refrain from talking to one another during a tasting session, and refrain from divulging their own results and experiences to any panelist that has not yet sat for the panel that day. Lastly, use this opportunity to establish the sensory group as a community of mutual respect. Take every chance you get to build a positive environment for your panelists because this will keep them coming back to sessions month after month, year after year.

Following the general introduction, you can move into teaching your trainees about flavor, their sensory systems, and proper beer assessment techniques. Refer to the material presented in chapters 2–5 to inform your discussion, but do not feel the need to cover each of these topics to the level of detail found in this book—you do not want to overwhelm your panelists with more than they need to know. The primary goal at this stage is to get them thinking critically about flavor and to teach them how to properly evaluate a beer.

In terms of specific topics, you should cover a general definition of flavor, highlighting the fact that flavor is influenced not only by taste and aroma sensations, but also by mouthfeel, appearance, and even sound in some cases. While on the topic of the different senses, you may wish to give trainees examples of various traits within the modalities of appearance, taste, and mouthfeel. You should pick and choose the examples you decide to cover based on the characteristics you feel are most important within your beers. In the context of appearance, you might show trainees examples of beers of different colors and clarity levels. For taste, you should discuss the different tastes perceived on the tongue and expose trainees to prepared samples of the common tastes (sweet, salty, sour, bitter, and umami), focusing trainees' attention specifically on the sour and bitter samples, as some people have difficulty distinguishing between these two tastes (see table 4.1, p. 50 for information on preparing taste standards). If desired, you can also present trainees with examples of different mouthfeel traits, such as body, carbonation level, astringency, and alcohol warmth.

Due to both its importance and its complexity, aroma should be covered in its own specific session. However, this initial session is a good time to cover orthonasal versus retronasal aroma. You can actively illustrate the retronasal pathway to your trainees using the jellybean test (see pp. 23–24), which will also demonstrate the significant role aroma plays in flavor perception.

Next, instruct trainees on the proper evaluation techniques for each of the sense modalities: appearance, aroma, taste, and mouthfeel. Have trainees perform each technique as you describe them to help cement the procedure in their minds. I recommend teaching the techniques using an unfamiliar beer to prevent your trainees from just going through the motions or leaning on preformed expectations of the flavors they will encounter. Although trainees may feel a bit uncomfortable with some of the techniques initially, reassure them by letting them know you will practice them together on each beer that you taste during training, and that the techniques will quickly become second nature.

Teach the techniques for each modality in the order you want your panelists to use them when assessing beers for a TTT test or a description test. I recommend using the order: appearance, aroma, taste, and mouthfeel. If you want to use a different order you should teach trainees that order from the outset.

You may wish to include a brief presentation of some of the most common biases present in sensory situations to help panelists watch out for these effects. In particular, you can reduce the impact of expectation error, selective perception, logical error, and the halo effect by teaching trainees how to recognize them when they arise. In addition, you can use the mutual suggestion bias as a good example to explain how biases affect our judgment, as nearly all tasters are familiar with this phenomenon. Bringing up the power of mutual suggestion will also reinforce the importance of not making reactive noises or talking to other panelists during a session.

Although you might not have time to cover test types in your first session, you should invest some of your training time in familiarizing panelists with the types of exercises you will ask them to perform. Typically, new panelists will primarily perform TTT tests, description tests, and (potentially) acceptance tests. Regardless of when you expose your panelists to different test types, make sure to give them adequate opportunities to practice each format—you do not ever want a panelist's first exposure to a new test format to occur during a live panel session. By allowing panelists to become comfortable with the test formats they will not have to think about the particulars of each exercise when they sit for a session and will instead be able to channel all of their focus into evaluating their samples.

Outside of general sensory knowledge, you may also wish to briefly cover other beer knowledge topics over the course of training. For example, covering the basics of the brewing process can be especially valuable if you have non-production employees on your panel. A better understanding of brewing ingredients and processes will help panelists contextualize the flavors that they pick up in each beer and can also lead to improved feedback in the event that a batch does fail a TTT test.

Brand Training

When you begin running sensory panels for product release, you will use the TTT test to assess outgoing batches of beer. To make sure you are getting reliable data with your TTT tests, you should familiarize your panelists with each of your brands. Brand training will form the foundation of your sensory quality control program. The more effort that you put into solidifying your panelists' brand knowledge, the stronger your sensory program will be.

Although the concept may seem obvious, you should begin brand training by explaining to panelists what defines a brand and how brand target descriptions are created. Emphasize that brand descriptions are not determined based on the ingredients used or the notes of the head brewer, but are, in fact, based on the observations of the sensory panel. If you already have brand descriptions in place, explain to trainees how those brand descriptions were developed by groups of existing panelists. If you are just getting started with your program, explain to trainees that they themselves will soon be developing the brand descriptions for your core beers. (For more information on developing brand descriptions, see p. 119.)

With a new sensory program, your panelists will need to develop an intimate understanding of your brands at the same time that they build the brand target profiles by performing description tests on multiple batches of each brand. If training a new group of panelists with targets already in place, begin brand training with a guided readthrough of the existing brand profiles. When performing TTT tests, panelists will evaluate each sense modality one at a time, so make sure to address the description for each modality separately. Follow this readthrough with a guided tasting of the brand in question, once again instructing trainees to focus carefully on each sense modality individually. This first tasting will allow trainees to internalize the key flavors and characteristics of the brand. If possible, allow trainees to taste a few different batches of a beer that all fall within the brand specifications. This will help them develop a feel for which characteristics tend to vary from batch to batch. It will also help them develop a feel for what constitutes an acceptable level of variation within a given brand. In sampling multiple batches, guide trainees' attention to any variations that you note between different batches, especially in these early stages while trainees are still honing their palates. Although not always possible, if you do have access to a recent batch of beer that was deemed not TTT, allowing trainees to taste that batch will help demonstrate what constitutes an unacceptable level of variation.

After familiarizing your panelists with your brands—either by guiding them through the process of building the brand profiles themselves or through exposure during training—use mock TTT testing to solidify your trainees' understanding of each brand. Initially, you should work with only one brand in a given flight. Present trainees with a group of samples (anywhere from three to eight), give them the supposed brand identity, and ask whether or not that sample is TTT. At this stage I recommend only asking for an overall TTT assessment rather than assessments of each individual modality, as this will allow you to work through a greater number of samples. Ask panelists to work through each sample one at a time, and not to compare the samples with one another. Supply trainees with the brand description to mimic the setup of a real panel session. Present trainees with a variety of different samples, at least half of which should actually be the brand in question. Other samples could include closely related brands that you produce, similar beers from other producers, or spiked versions of the indicated brand.

Testing trainees in this way helps them establish a feel for the brand in a blind tasting setting. Cognitively, the task of identifying flavors in a known beer is drastically different (and significantly easier) than trying to identify flavors in a beer of unknown identity. This type of testing closely resembles the sort of assessment that panelists will have to make when they sit down to assess a batch of beer during a TTT test, and exposure to the format will make them more comfortable with the task when it comes time to perform the test for real. If you have time, you can administer these sorts of tests across multiple training sessions to really ingrain each brand's identity in your trainees' minds. Once you have conducted training on multiple brands and your trainees have grown familiar with the test format, you can begin to administer test panels that involve multiple brands to simulate the conditions of a real product release panel (e.g., one TTT test on your American IPA, one

TTT test on your German-style Pilsner, and one TTT test on your German-style weissbier).

WHAT ABOUT ONE-OFF BEERS?

In the current brewing landscape, many breweries produce single batches of experimental beers with no intention of ever producing these beers again. Some breweries take this model to its logical endpoint and do not have core brands of any sort, with each new beer representing a one-time creation. Does this mean that these breweries do not need to taste these beers before they go out to trade, or that true-to-target (TTT) testing becomes irrelevant? Absolutely not! However, if your brewery fits this model, brand training will play a significantly reduced role in your panelist training program. Instead, you should monitor beers for product release using a hybrid approach involving a description test followed by a TTT test, mirroring the recommended process for building a new brand. For full details on this method, see Defining and Shaping a New Brand, p. 143.

Flavor Training

Beyond general knowledge, tasting techniques, and brand training, any additional training you give your panelists is icing on the cake. Equipping them with a lexical tool like the Beer Flavor Map should be sufficient to get them started on building brand targets —you will be amazed at how quickly your panelists align around specific flavor descriptors. Further investment in flavor training or attribute training will boost your panelists' abilities on description tests and TTT exercises but are not necessary to get your program off the ground. If you have the time during initial panelist training, consider pursuing some of the following topics. If not, you can always introduce this material after you have started running regular panel sessions.

Aroma Training

After a few rounds of description testing with your new panelists, you will find that they gravitate to a subset of the descriptors presented in the Beer Flavor Map (or whatever lexical reference you decide to use). The flavor terms that trainees use the most are prime candidates for aroma training, which can help ensure they use flavor descriptors accurately. Alternatively, if you already have brand targets developed for your beers, the flavor descriptors used in those profiles make a logical group for your first round of flavor training.

As mentioned in chapter 4, you can use grocery standards (e.g., using fresh grapefruit segments to train panelists on grapefruit aroma) to both strengthen your trainees' ability to recognize flavors and help the group coalesce around a specific flavor profile for each descriptive word used. To perform training with grocery standards, build "flavor pots" for each standard. A flavor pot simply consists of an opaque container with a lid (an odorless paper or plastic cup covered with a petri dish will work) filled with some amount of the substance in question. You want your trainees to learn to identify the standard by aroma alone, and to divorce their mental image of the standard from other cues like appearance, taste, or mouthfeel. When a taster describes grapefruit flavor in beer, they are not usually referring to the taste or texture of a grapefruit, but rather to its aroma. To ensure that trainees map their flavor vocabulary to the aromas rather than any of the other characteristics of the item, instruct them to only smell the flavor pots; trainees must avoid looking in them, shaking them, or tasting the contents. Remind the trainees that the goal of these exercises is to learn, not to be right. If they look in the pots to come up with the correct answer, they are only cheating themselves.

In addition to grocery items, you can also place actual beer ingredients, such as crushed malt or raw hops, in flavor pots to simulate the flavors of these ingredients. However, use caution and avoid placing too great an emphasis on actual brewing materials, as the flavors of raw malt and hops do not always translate to the flavors found in a finished beer. Training panelists to accurately recognize, say, the smell of Cascade hop pellets may not offer much utility for your panel.

To train panelists on grocery standards, use recognition training methods (see p. 48). While you may be tempted to cover a wide range of flavors right off the bat, avoid trying to build a large vocabulary too quickly. You may have 20 or 40 different flavors that you want your panelists to know, but starting with such a large group can result in trainees losing confidence in their ability to consistently identify any of the desired flavors. Begin by focusing on the most common flavors found in your beers. Once your trainees grow comfortable with a small group of flavors, you can begin to expand their lexicon.

Taste Training

Taste training can go beyond what is covered in general knowledge and skills. Depending on the needs of your program, you may decide that trainees do not need additional taste training beyond your initial introduction of the basic tastes. However, if you want to further refine your trainees' ability to scale tastes, you can set up subsequent rounds of training, focusing on the tastes that are most important in your beers. If you were to only train on one of the tastes, bitterness makes for a good candidate since it defines the balance of innumerable styles. Training on sour taste will serve you well if your brewery produces any beers with notable acidity, but it can also be useful to breweries in general because it helps panelists spot a nascent bacterial infection. Outside of these two tastes, you may find value in training panelists on sweetness levels, or perhaps saltiness levels if you produce a Gose as one of your core beers.

Rather than training new panelists on differing levels of a taste attribute in water, have your trainees practice any additional taste training in a beer matrix. Unlike spiking with flavor attributes—most of which are not usually present in the base beer—you will have to adjust the amount of tastant added depending on the how much of that taste already exists in the beer. For example, if training with an American wheat beer that normally measures 20 IBUs, an increase of 10 IBUs should be enough to produce a noticeable change in perceived bitterness. However, if you attempt the same adjustment on your imperial stout that is already at 85 IBUs the difference may be imperceptible.

To effectively train on differing levels of a given taste, select a base beer low in that particular taste and then spike to a medium and a high level, which you can determine through trial and error. Alternatively, you can select three beers that you feel accurately represent each level of bitterness (or acidity, or whatever taste you are after), and train your panelists using these beers as reference points along the scale. While certainly less work than experimenting with and preparing samples, this method has a few disadvantages. By using existing beers, you will not have the chance to dial in the level of the taste to the exact amount you desire. And since trainees will taste three potentially very different beers, they may have trouble isolating the specific taste because the beers may differ across many other attributes. However, given the relative importance of taste training versus flavor training or brand training, this may be sufficient for your panel. Remember, you have a limited amount of time to train your panelists—any time spent training for one skill set takes time away from training for another. Focus first on the capabilities most important to your panel's function. You can always train on other skills once you have your panel up and running.

Attribute Training

If you have the means to do so, I recommend training your panelists on a core group of specific beer attributes as well. As with flavor training, begin by training new panelists on a limited set of attributes to allow them to master recognition of the compounds presented before trying to expand to a wider group, and limit the flavors you choose to attributes that will serve your panelists during panel sessions. Dismiss the notion that you cannot have a functioning panel unless your panelists can consistently recognize a group of 10 or more attributes—this places an unnecessary hurdle in the path to assembling and running your panel.

You can conduct attribute training in the same time frame or even the same training session as basic flavor training, but you should not mix grocery standards and specific attributes within a single recognition panel. Before beginning attribute training, thoughtfully select a small group of attributes that you would like your trainees to master. I strongly recommend including diacetyl and acetaldehyde on your short list, and add isovaleric acid if you produce hoppy styles of beer. If you have a barrel program or produce any sour beers, acetic acid should also make the cut. Beyond this core group of three or four attributes you can add other compounds, such as those listed in chapter 5 (pp. 60–67), but I would encourage you to keep the initial group small. This will keep training and supply costs down and will also help panelists build experience and confidence on the most essential attributes that may show up in your beers. Ultimately, the attributes you choose should meet the needs of your brewery. If your brewery has experienced a periodic or persistent flavor issue not covered here, add it to your personal short list of essential attributes.

To train panelists on specific attributes, use the standard recognition panel format presented in chapter 4 (p. 48). Developing proficiency in attribute identification requires repetition. While important in all types of training, repetition matters most with attributes, as the aromas of specific attributes will often be unfamiliar to trainees upon initial exposure. Beyond initial training, you can give panelists further exposure to these attributes through additional training or simple training stations, which I will discuss in the next section.

PANELIST MAINTENANCE

Panelist training does not end when panel sessions begin. If anything, their sensory journey has just begun at that point. Regular panel participation will help your tasters stay sharp, but additional training will ensure that your panel's accuracy remains stable, or even improves, over time.

Upkeep Training

Upkeep training should balance two distinct but related goals: maintaining your panelists' current knowledge base and performance, and improving and expanding your panelists' knowledge base. Over time, expanding your panel's descriptive lexicon and increasing the range of attributes that your panelists can identify will improve the data produced by your panel, but only if it does not come at the expense of practicing on the flavors and attributes that they already know. Sensory skills, like all performance activities, require maintenance. If a distance runner trained to run a marathon and successfully completed their goal, you would not automatically assume that they would be able to repeat the same achievement without continued practice. If a runner allowed several years to pass without keeping up their training, their ability and performance would inevitably decline.

This same phenomenon occurs in the world of sensory performance. A panelist may receive 10 exposures to diacetyl over a two-week period and may then demonstrate perfect recognition of the compound in a series of tests. However, if the panelist goes a year without further exposure to diacetyl, their detection threshold will increase and their ability to

identify the compound will decrease. I do not mean to discourage you from exposing your panelists to new flavors or attributes, but rather to remind you that, as humans, we have a tendency to learn a topic, assume that we have mastered the material, and then move on. Make sure that you re-expose panelists to the flavors and attributes that they have learned, either by periodically spiking samples in a panel session or by offering additional training in another form. If you find panelists struggling with a flavor that they previously learned (e.g., several panelists miss an acetaldehyde-spiked sampled in a TTT test), you should try to find ways to give your panelists additional exposures to that flavor to help them relearn its characteristics.

Some panel leaders approach continuing education by offering ongoing training sessions, usually on a weekly, biweekly, or monthly basis. If you have the time to organize and administer regular structured trainings (and your panelists have time within their own schedules to attend these trainings), this will certainly help expand the skill set of your panelists. However, with a bit of creativity, you can create other types of exercises to help panelists stay sharp. Consider moving the exercises outside the confines of typical classroom-style instruction by setting up stations with simple exercises that only take a few minutes to complete.

Stations can serve a variety of goals depending on how they are set up. To improve flavor recognition, a station can consist of several different flavor pots of grocery standards for panelists to attempt to identify. To boost attribute recognition, a station can contain a pitcher of control beer next to a few pitchers of unlabeled spiked samples that panelists can test themselves on. A station can even focus on brand training by presenting panelists with a description of a known brand and then three different pitchers to sample from, some containing the brand in question with others containing a similar beer. In all cases, these stations involve presenting samples to panelists in a self-administered fashion while obscuring the identity of the samples so that panelists can still blind test themselves to help build their skills. You can also create familiarization stations to present panelists with new flavors or attributes, followed by a blind testing station on those same flavors the next week. Focus on blind tasting activities whenever possible, as blind tasting best mimics the conditions of a panel session and helps panelists strengthen their sensory abilities within that context. Regardless of the type of station, the activity should be something that a panelist can complete within five minutes. Keeping the activities short makes it easier for panelists to participate, increasing the likelihood they will actually engage with your stations.

To collect panelist responses from these stations, use some sort of digital method, such as a sensory app or a Google form attached to a QR code. This allows you to rapidly build ballots, change sessions on the fly, and easily compile data from the test, while also giving you a mechanism to instantly communicate the identity of the samples to panelists once they have entered their results. This way, a panelist can interact with a sample a second time once they know its identity, which will allow them to further cement their knowledge and to correct any mistakes they made.

The station approach offers much more flexibility compared with full training sessions because you do not have to be present at the station while panelists interact with the samples. And since panelists can visit the station at any point in a wide time window, they can work it into their own schedule, making it easier for them to participate. Additionally, in terms of learning a skill, short, frequent practice sessions are much more effective than longer, infrequent sessions. You will have much greater success in training your panelists if you can get them to practice a little bit each week rather than having one long training session each quarter. Even if you do conduct regular ongoing training, these sorts of stations supplement the lessons you teach in class by offering additional opportunities for practice.

As you think through ways to keep your panelists learning and engaged, do not allow the suggestions here to limit your approach. Try to think of ways to infuse fun, and perhaps even a bit of lighthearted competition, into your trainings. As an example, Karl Arnberg of Allagash Brewing Company related a sensory-inspired riff on the card game memory that he likes to play with his panelists. In the original game, picture cards (containing a variety of images, such as a banana or a rabbit) are laid face down to form a grid, with two of each image present on the board. On their turn, players flip over two cards in an attempt to form a pair, returning the cards to the board face down if the cards do not match. At Allagash, Karl sets up a game board with a six-by-six grid of samples. He pours two pairs each of seven spiked beers into various cups around the board and then fills the rest of the cups with the control beer. On each panelist's turn, they are allowed to smell two samples, attempting to find a match. Panelists can then communicate their impression of their two samples to the other members of the group. Sometimes Karl will pit panelists against one another to see who can collect the greatest number of pairs, while other times he will have the group work collaboratively to see how quickly they can clear the board.[1] Novel approaches to training such as this can help keep panelists engaged and excited about their participation in your sensory program.

Maintaining Panelist Motivation

Given that many brewery employees work in this industry due in part to their love of beer, you would think that you would never have trouble convincing your coworkers to drink beer as part of their job. However, ask anyone who has led a sensory panel in a brewery—or in any industry for that matter—and they will tell you that getting panelists to show up week in, week out is one of the greatest challenges in running a successful sensory panel. Part of this owes to the fact that, as you can probably see, sitting on a sensory panel does not bear much resemblance to drinking beer for enjoyment. While panelists probably do not consider sensory work physically arduous, it still requires great focus and mental effort, which squarely differentiates the experience from enjoying a relaxing pint at the end of a shift.

Some breweries will use a variety of different rewards to help encourage panelist attendance. Most sensory leaders serve snacks to panelists following sessions. In addition to serving as an incentive, this form of reward can also serve as a visual cue to panelists who have not attended yet that day. When they see a fellow panelist enjoying a post-panel snack it can serve as an enticing reminder to go and participate on that day's panel. Some breweries decide to offer gift cards, merchandise, or pay bonuses, either to the panelist who attends the most sessions or to all panelists who attend a certain number of sessions within a given time period. Custom apparel available only to panel

---

[1]   K. Arnberg, telephone conversation with author, April 10, 2020.

members can make panelists feel like part of a team, allowing them to show off the pride they feel, and can also encourage other employees to join. While these tactics help to recognize panelists for their hard work, they do not address the root causes that might be driving lack of panelist enthusiasm.

In some cases, sagging panelist morale can indicate underlying company culture issues. If your brewery cultivates a positive and engaged workplace where employees care about their company, you will find it easier to recruit excited, motivated panelists. If problems with panelist engagement still plague the sensory program in such an environment, it may stem from a lack of leadership buy-in. For a sensory program to succeed, the leaders of the company cannot just say that sensory matters, they have to show employees through their actions. This means managers actively encouraging their direct reports to participate. This means directors not scheduling meetings or events that conflict with sensory panels. If a panelist feels that their manager disapproves of their panel participation, this will serve as a powerful demotivator. If you can build a culture that appreciates the value of sensory, you should have no trouble attracting panelists.

Retain panelists by paying attention to their needs and wants. Some common motivators for panelists include continuing education, self-improvement, and the challenge they find in sensory work. Give panelists feedback on their performance, noting their strengths as well as areas where there is room for improvement. In addition to providing panelists insight into their own tasting abilities, this shows them you are actually paying attention to their individual performance and trying to help them grow as panelists. Schedule quarterly one-on-one meetings with each panelist to allow them to interface with you directly in a private setting. Solicit feedback on how you can make the program better and listen to what they say. Show your panelists that you respect them and that you appreciate the hard work they put in and they will return the favor through their continued attendance.

Finally, and perhaps most importantly, make sure that your panelists see the results of their hard work. The most successful sensory programs that I have observed were the ones that reinforced the importance of the program by consistently and proactively communicating—both to managers and panelists—the ways in which data collected by the sensory panel affected decisions made within the brewery. Depending upon how often you conduct panels, this may be a weekly, monthly, or quarterly communication to panelists and managers. If you can show panelists that their participation in a panel prevented a flawed beer from reaching the market, helped shape a new brand, or led to a process improvement within the brewery, your panelists will take pride in their involvement. Show them that their attendance impacts the beers produced by the brewery. The fulfillment that this knowledge brings them will serve as a stronger motivator than any snack or reward ever could. I am not arguing that you should do away with traditional reward systems, but your first priority should be to show panelists that their work matters.

10
PRODUCT RELEASE PANELS

Once your new panelists have undergone a bit of training, you can begin running sensory panels to answer questions about your beers. The remainder of this book deals with a number of topics you can explore using sensory, beginning with our primary focus—quality control and product release. To some people, the idea of sensory quality control entails panelists carefully checking each batch of beer for flaws, with the goal of preventing bad beer from reaching consumers. And while the test you use for quality control should prevent that outcome, we want a test that can do more than simply identify problematic beers. A strong quality control program does not just prevent the release of bad beer but guarantees the release of good beer. True-to-target (TTT) testing serves us well in this capacity by asking panelists to check each beer against a target profile. Using this method, panelists will check to see that each beer exhibits the flavors and traits that should be present while also searching for any unusual flavors that should not be there. By using TTT testing to assess every outgoing batch of beer, you ensure the beers you sell actually taste the way you want them to.

DEVELOPING BRAND TARGETS

As you will see, TTT testing has a wide variety of applications in sensory work. However, regardless of what you use the TTT test for, the first step is always the same: establishing the targets against which you will test samples. In the case of product release, you will work with your panelists to build brand targets for your finished beers.

Use the description test (p. 87) to develop your brand targets. If you are just getting started with your panel, you should provide your new panelists with a lexical tool like the Beer Flavor Map before trying to have them define brand profiles for your beers. Tasting, like other skills, requires practice, and your new panelists will get better at descriptive profiling over time. Before tackling your own brand profiles, you may want to build up your panelists' experience with a lower-stakes exercise like creating a brand description for a beer from another brewery. By reviewing the results of these practice description tests with your panelists, you can help them shape and refine their shared lexicon. Do not go overboard here in an attempt to make sure that your target profiles are perfect—a few rounds of practice description tests should suffice. Once your panelists grow comfortable with the description test, you can move on to defining targets for your brands.

Good brand targets will deconstruct the description of a beer across each of the sense modalities. Refer to chapter 7, page 85 to determine the exact list of attributes you want to track within each modality. Your target profiles should clearly address each attribute measured, so remind panelists to evaluate each of these attributes every time they perform a description test. For an example of a brand profile, see figure 10.1.

Your brand targets should be representative of the beer's key characteristics, but permissive enough to allow for a small amount of batch-to-batch variation. To make sure that your targets accurately describe

Brand target:	Belgian Witbier
Appearance:	Pale straw color, medium haze, white foam, high foam retention
Aroma:	Coriander, lemongrass, and orange peel, with slight black peppercorn, white bread, and grainy aromas
Taste:	Low bitterness, low sweetness, no sourness
Mouthfeel:	Medium body, high carbonation, no astringency, no alcohol warmth
Overall:	Must feature pale straw color, medium haze, coriander aroma, low bitterness, and high carbonation. Cannot contain perceivable sourness or vegetal celery notes

Figure 10.1. An example brand target for a Belgian-style witbier.

the brands they represent, build each target using description tests taken across three to five different batches of that brand. Once you have performed a few description tests, average the results to build the final target profile. This will ensure that your target description captures the range of characteristics possible for that brand. After compiling the descriptions to build the composite target, you can also add an "overall" category to the description. Panelists will use this description when making their overall determination on whether a sample is TTT, so it should capture the key flavors that must be present in order for the beer to pass. Also, if a particular brand sometimes develops certain unacceptable flavors, you can list those in the overall description as characteristics that must *not* be present (fig. 10.1).

When performing each description test, do not give panelists clues or hints of the beer's identity. To build a truly representative profile, panelists should evaluate the brand each time as if it were their first time tasting it. Tasting different batches of each brand will necessarily space the tests out, which should prevent panelists from simply repeating the responses they remember giving during their previous assessment. You want panelists to use their senses to describe what they perceive, rather than what they can remember from the last time.

Before finalizing your brand targets, be sure to get management to buy in and sign off on the descriptions. This can help mitigate the potential for confusion or misunderstanding down the road once you begin using the profiles for TTT testing. Finally, present the finished target descriptions to your panel and train your panelists using the methods outlined in Brand Training, p. 112.

TESTING PACKAGED AND IN-PROCESS BEER

With your established brand targets in hand, you can begin holding product release panels to monitor your beers. For a product release panel, panelists will use the TTT test to evaluate each sample that you serve to them (see p. 73 for TTT testing format). At the outset of a session, present panelists with all of their samples at once, assuming that you have enough room to do so. If space is limited, you can serve samples in multiple rounds. Order the samples based on their intensity, with beers that have elevated alcohol and bitterness levels or just generally have a more intense flavor profile coming later in the panel. Instruct panelists to work through their samples in the order presented, and to only work on one beer at a time. I suggest limiting the number of TTT samples in a single session to eight, though you can go as high as 10 samples with more seasoned panelists. With high-intensity samples, you may want to consider limiting the number of samples served even further. Beers high in acid, alcohol, bitterness, or other attributes can quickly lead to palate fatigue.

The TTT test is quite malleable and can be used to monitor multiple process points within the brewery. When just beginning your sensory quality control program, you should focus on monitoring each batch of freshly packaged beer. When assessing a given batch of beer, your panel should test each package type used for that batch. For example, if a batch is split between bottles, cans, and kegs, at least one bottled sample, one canned sample, and one kegged sample should show up on your product release panel. By monitoring freshly packaged beer, your panel will taste each batch of beer immediately before it goes out to the trade. This is the best way to ensure that your brewing process up to and including packaging is free from defect.

Beyond giving you data on individual batches, monitoring packaged beer will also allow you to spot emerging trends in your data. This can help you catch any sort of drift in the characteristics of a given brand. If a brand begins to drift, you should bring this information to the attention of the production team, allowing them to bring any errant brewing processes back in line. Evaluating beer at this stage will also allow you to prevent truly flawed beer from leaving the brewery. However, catching a borderline beer at this stage does not offer much in the way of recourse. Additionally, each step through the brewing process, from raw ingredients to wort to beer and eventually packaged beer, increases the cost of the final product from the perspective of time and resources invested. Packaging materials in particular often constitute a significant portion of the raw materials cost, so anything you can do to prevent flawed beer from making it into package will help improve your bottom line. To this end, you should also monitor bright tank beer prior to packaging.

Assessing Bright Tank Beer with TTT Spot Checks

Monitoring bright tank beer offers several distinct benefits. First, it allows for a quality check before the beer goes into package. This means that if a truly flawed batch of beer makes it this far, it can go down the drain rather than into packages that will ultimately need to be destroyed. Depending upon the size of your program and the frequency with which you make each brand of beer, catching a minor flaw in the bright tank might allow you to blend that batch with another, mitigating the issue entirely, whereas that option is not really available to you if you catch an issue in a beer post-packaging. However, sending a bright tank sample from each batch to your panel for TTT testing is often impractical, particularly for smaller breweries that may only hold sensory sessions once or twice a week. If finished batches of beer end up stuck in bright tanks waiting for sensory panel sign-off before packaging, the rest of the brewery may come to resent the sensory program for holding up production. On the flip side, packaged beer can be held for several days before going out to the trade to allow for sensory testing without inconveniencing the core operations of the brewery. Nonetheless, you should still assess the beer in bright tanks in some way prior to packaging.

To accommodate these opposing goals, you can use spot checks—a modified TTT test of sorts—on each bright tank immediately before it goes to package. The spot check should consist of simple criteria to evaluate and should use a streamlined version of the full brand profile. While you do not need a large group of panelists to taste each tank, at least two people should sign off on the beer before packaging begins.

The first criteria for the spot check should simply be a visual test: does the beer exhibit the correct color and correct level of clarity for this brand? Basically, this ensures that you are, in fact, packaging the correct beer. While it may seem like a trivial check, I have

heard horror stories from breweries large and small in which the wrong bright tank was hooked up to the packaging line. When the brand on the packaging does not match the liquid in the bottle, your sole recourse is to destroy the batch. This simple visual assessment prior to packaging can serve as a quick check against this type of error.

The spot check should also incorporate a basic evaluation of the flavor profile of the beer. Because the spot check occurs under non-ideal conditions—on the brewery floor by a small number of potentially untrained tasters—you should not expect the tasters to pick up on nuanced elements of the beer's profile. In most cases, you can simply use the overall section of the brand's full profile as the target for this test, as that section should encapsulate the most prominent flavors of the brand. During a spot check, tasters should confirm the presence of these desired flavors and also quickly check for any unexpected flavors such as diacetyl or acetaldehyde. Assuming that both tasters sign off on the beer, the beer moves on to be packaged, and receives a more thorough assessment from the sensory panel before going out to the trade. While this spot check will not catch every single issue, it can help greatly mitigate risk. By keeping the TTT spot check simple and broad, brewers or packaging employees can perform it quickly without slowing down production.

Monitoring Additional Process Points

In addition to bright tank and packaged beer, you may eventually wish to use TTT testing to assess other brewing process points as your program grows. Some breweries will monitor their fermentors, typically as the beer nears the end of maturation. Once again, if you can catch an issue further back in the process, it both increases the likelihood that you will be able to fix the issue and reduces the resources invested in the beer if you end up having to dump it. However, several panel leaders commented anecdotally that issues or anomalies that come up during fermentation will often sort themselves out during maturation. If you do perform TTT testing on fermentor samples, you also need to develop corresponding brand profiles by using the description test on fermentor samples. As you probably already know, beer flavor changes significantly between the fermentor and the finished package. Panelists should not have to evaluate a fermentor sample based on the description of the finished beer.

The decision to increase the number of process points monitored should not be made lightly, and in many cases the cost of monitoring additional parts of the process outweighs the benefit. Do not underestimate the time required to add another process point to your assessment of each batch. In addition to the initial time investment to create fermentor brand profiles and train panelists on a second set of profiles for all of your core beers, also consider that you will effectively double the number of samples that your panel needs to taste for each batch of beer. Depending on the number of panelists that you have, this may end up stretching your panelists too thin. Be careful, as oversampling in this manner can eventually lead to panelist burnout.

Maintaining Panelist Focus

Over time, TTT testing will become repetitive for panelists. Assuming that your production team has their process dialed in, panelists may get in the habit of passing beers through as TTT. Though most panelists are conscientious and do not intend to taste samples in an unfocused way, biases such as the default effect can cause panelists to miss small changes in beers over time. To combat these biases, keep panelists on their toes by occasionally spiking TTT samples in your product release panels. If panelists know any given sample might contain a spike, they will pay more attention across the board to avoid missing the spiked sample when it does eventually turn up during a panel. Do not set any sort of regular schedule for spiking samples, as you do not want the introduction of spiked samples to become another predictable feature of your TTT panels. The exact spiking frequency you choose will depend on how often you hold panel sessions, but most panel leaders recommend somewhere between once a week and once a month.

PRODUCT RELEASE ACTION STANDARDS

As discussed in chapter 6 (pp. 74–76), you will analyze data from your product release panels using p-charts (or using a discussion-based format following individual tasting if you have less than eight panelists). Most samples will be marked not TTT by at least one panelist, but that is usually not a cause for concern. The data analysis methods you use will allow you to see when a batch actually requires further attention. So, what do you do when a batch fails a TTT test? Do

you sound the alarm? Do you have to dump the batch or can you still sell the beer?

Unfortunately, I cannot give you a simple, straightforward answer of what needs to be done when a beer fails a TTT test. Your exact response will vary depending on the situation, and your potential remedies will be determined by both the issue at hand and the capabilities of your organization. However, I can give you some general guidelines and best practices for determining your response to a failed TTT test. You should adapt the following considerations to create action standards that work for your brewery and update them over time as you gain more experience.

ACTION STANDARDS FOR BRIGHT TANK SPOT CHECKS

Just as you should have action standards for your formal product release panel TTT testing, you should have standards in place that dictate what to do if a bright tank spot check fails. A spot check entails a simpler test than a full-on TTT assessment, and the corresponding decision tree is simpler as well. First and foremost, if a beer fails the visual inspection of color and clarity, do not package that beer! In this case, the packaging team obviously should not proceed until they have selected the correct bright tank. If the flavor inspection fails, the tasters should bring the sample to you for further testing. If you have time, you can run the sample through your standard product release panel. If that is not possible, you should still have a few of your panelists taste the sample to determine whether it is fit to package. Next steps will usually involve meeting with the production team to discuss whether they can do anything to remedy or mitigate the issue, whatever it may be.

Verify Results

When a sample fails a TTT test, you first want to confirm that you are reacting to an issue that actually exists. Remember, there is always the possibility that sampling error or even random chance will cause panelists to fail an acceptable sample. Preparation mistakes or judgment errors in the lab can lead to a panel failing a sample. As illustrated in chapter 3 (Adaptation Effects, p. 34), even sample order effects can sometimes sink a sample.

Begin by reviewing panelist feedback to determine whether the panelists who failed the sample gave similar reasons for the decision. If multiple panelists give the same sort of feedback (e.g., "bitterness is too low" or "DMS present"), you may be able to identify a specific issue right off the bat, which will inform your next steps. If feedback from panelists who failed the sample does not coalesce around a single cause, you should run the sample again at one of the next panel sessions. When retesting a sample, avoid letting panelists know that they have already tasted the sample previously—this follows standard practice, as samples should always be presented without any identifying information. However, you also want to make sure that panelists do not find out that a sample of that brand previously failed panel, as this information will heighten panelists' suspicions and bias their results. The goal in retesting a sample is not to run the sample over and over until it passes panel. A sample that truly falls outside of brand specifications should fail upon a second panel evaluation, thus, running the sample through your panel a second time greatly diminishes the risk of a type I error (false positive).

Even if your panelists' feedback does appear to indicate a specific cause, you may still want to examine the conditions of the test and consider running the sample again anyways. Certain errors in the lab can produce misleading or misrepresentative data. Contrast effects from sample to sample can skew panelist performance. In some cases, simply changing the order of samples can yield entirely different results. Be cognizant of the fact that your lab practices can potentially introduce errors into your data.

If your retest of a beer that has previously failed comes back with a p-value close to the panel average, you can probably consider the initial result a fluke. However, you should preserve the data showing the sample initially failed panel and keep an eye on that batch over time. Try to correlate the data produced by your sensory program with general market performance, particularly if you release any batches that did not perform well in a product release test. Note whether any consumer complaints come in, but do not automatically assume that a lack of consumer complaints means the beer was totally fine. Collecting

and tracking this data over time will help you make better decisions if a similar issue arises in the future.

What Next If a Batch Has Failed?

Once you have determined that a batch is not true-to-target, you have to answer two questions: what do you do with this batch of beer, and what happened in the brewery that caused the batch to turn out this way? Ultimately, determining what to do with a batch of beer that fails a product release panel will involve a conversation with management or key stakeholders at the brewery. To best inform that discussion, you should first try to figure out what went wrong, as that information will affect how the brewery decides to move forward.

Begin your investigative work by using your panelists' feedback to try to diagnose the issue. Sometimes a distinct flavor issue, such as diacetyl or isovaleric acid, will point you in the direction of a cause. If you are unable to ascertain the issue based on panelist feedback, meet with the production team to review brewing data for that batch to see if anything anomalous occurred during the process. Sometimes bringing the issue to a broader group can help you get to the bottom of the issue, as one of the brewers or a member of the packaging team may be able to point to something weird that happened with that batch. Assuming that you are able to identify the root cause, you can work with the production team to establish practices that prevent the issue from occurring in the future. If you are unable to diagnose the problem, keep a vigilant eye on samples coming through sensory panel to watch for similar issues in other batches or to see if any pattern emerges. Sometimes, you will spot trends that allow you to solve the mystery (e.g., "we get this weird flavor in our beers with seventh-generation yeast").

The decision of what to do with a flagged batch of beer falls to brewery management or leadership. As a best practice, the sensory department should not serve as a decision maker, particularly when the decision has explicit financial implications for the brewery. Instead, the sensory program should provide data that support the decision makers. Separating decision-making responsibility from the sensory department allows you to report exactly what you discover without having to qualify your findings to support a particular outcome. If you know that presenting a not TTT finding automatically results in a dump decision, you may hesitate to present borderline cases. Removing decision-making authority allows you to steer clear of emotional responses and focus on your core duties—analyzing and tracking the quality of the beer—without having to worry about how your data may impact brewery operations.

To prepare for the discussion of a flawed batch of beer, you should establish a set of questions to cover. Going into the conversation, you ideally should be able to answer the following:

- **How big of a deal is this?**
  Does the batch have a clear, identifiable issue, or did the sample fail without panelists being able to clearly identify a problem? How far off is it from the brand profile? If we sold the beer, what proportion of consumers do you think would notice the issue?

- **How do we expect this problem to age?**
  Is this an issue that will get worse over time? Is this an issue that will potentially clear up over time?

- **Have we seen this issue before? If so, what did we do then? Did we get consumer complaints?**
  At the outset of your sensory program, you obviously will not be able to refer back to previous instances where an issue has come up, but over time you will develop an understanding of how certain types of issues play out in the market. Of all the questions on this list, this one perhaps best informs how to proceed with the batch in question.

- **Do we know the root cause, and have we identified the steps to prevent this from happening in the future?**
  This information will be the priority once a decision has been made on how to deal with the flagged batch of beer. If you cannot get to the bottom of the issue prior to this meeting, at least try to eliminate some possibilities so that you can have a constructive discussion about how to potentially improve in the future.

The answers to these questions will likely determine the next steps for that batch of beer. But bear in mind that there are options at your disposal aside from dumping the beer. If the beer shows flaws that you think will get worse over time but are not yet bad enough to preclude its sale, you may decide to keep the beer in-house and only sell it through your taproom. Similarly, you may decide to send the beer to a market where you expect it to move quickly, or you may keep it confined to local markets where it will be easier to pull back if the issue gets worse. Any time that the brewery decides to sell a borderline beer, your panel should continue to monitor samples from that batch to determine if and when to pull it from the market. The decision to dump a batch of beer carries financial consequences and should not be made lightly. However, if the beer presents notable flaws, destroying that batch may be the only way to preserve the integrity of your brewery's brand and avoid the significant consequences of having to deal with an avalanche of customer complaints or a product recall.

In any case, communicate openly with your panelists about what happens to beers that fail a product release panel, even if the brewery ultimately decides to sell it. Without proper framing, your panelists may find this disheartening or feel like their work does not matter. You should explain to them that their input did inform the decision-making process. You can also ask leadership to help deliver the message and explain what led them to release the beer anyway—direct interaction between key stakeholders and panel members helps reinforce the idea that the company heads value the panelists' contributions. There is merit in capturing data about an off-target batch of beer and making the decision to sell it regardless, because you can track how it performs in the market and use that information if you encounter a similar issue in the future. Ultimately, your brewery will need to determine how much variation it is willing to accept in its beers. Sensory quality control is all about mitigating risk and using the data you collect to improve in the future. In addition to providing feedback to the production team, product release results correlated with market performance data will help refine the decision-making process behind when to release and when to dump a borderline batch of beer.

11
BEYOND BEER RELEASE

With your panelists trained and product release testing in place, you have successfully established the core of a sensory-based quality control program. Congratulations! This is no small feat, and it will serve your brewery well for years to come. Now you can begin to consider other ways in which your sensory program can be used to improve operations at your brewery. While you have several options that you might pursue, I recommend first tackling ongoing quality control for beer that has already entered the market.

If you sell the majority of your beer on draught at your own taproom, most issues relating to shelf life will not apply. However, if you sell packaged beer to go—and especially if you distribute packaged beer—managing beer that has left the brewery becomes an important and highly complex challenge. In this context, your sensory program can help address two key issues: how to best respond to consumer complaints and how to determine an appropriate shelf life for each of your beers.

MAINTAINING A BEER LIBRARY

In addition to monitoring each batch of beer before it leaves the brewery, you should also retain a small amount of beer from each batch in case future testing becomes necessary. Building a beer library allows you to refer back to specific batches of beer should a customer complaint arise. In building a beer library, you need to decide how much beer to hold, under what conditions to store the beer, and how long to

hold the beer for. The size and capabilities of your brewery will likely dictate the answers to some of these questions. Anecdotally, a large packaging brewery might choose to store one case under refrigeration and one six-pack at ambient temperature for each batch produced. However, for smaller breweries, storing that amount of beer may not be feasible or even desirable. Here I will break out my suggestions into three tiers—your goal should be to eventually reach tier three, but you can begin at tier one and grow your library as your program evolves.

BATCH LABELING

In order to track beer in the market, you must have a way to identify a specific batch of beer based on the packaging. Most breweries accomplish this using some sort of batch coding. If you do not already label your beer with a batch code, you should begin doing so. This is best practice, not just from the perspective of sensory follow-up but from a health and safety standpoint as well. If your brewery ever has to issue a recall on a specific batch, doing so will be virtually impossible if you cannot determine the batch from the product's packaging. Without such information on the label, a beer recall will require pulling every package of the affected brand from the market, rather than just the batch in question.

At the first tier, you do not actually need to hold on to any beer at all. This tier simply requires that you follow good batch labeling practices and that you perform product release TTT testing on each batch of beer before it leaves the brewery (which you should already be doing at this stage!). As long as you do both of these things, you will be able to refer back to your sensory data for a given batch if a customer complaint comes in. In reviewing your original data, you may find that the batch was perfectly normal, or you may find that a higher-than-normal percentage of panelists marked the sample as not TTT. Either way, correlating a complaint to your product release testing data can help you track down whether the issue occurred before or after the beer left the brewery.

Additionally, in responding to a customer complaint, always ask the customer if they would be willing to ship any remaining beer to you (obviously, offer to cover shipping and reimburse them for the beer). If you do not have the space to maintain a robust beer library, obtaining a sample of the beer that generated the complaint will give you the ability to run follow-up testing based on the customer's comments. Even if you do have samples of beer from that batch in your library, soliciting the affected package from the customer offers several advantages. First, it allows you to test the customer's specific package to determine whether a flaw actually exists. Second, you can potentially determine the nature of the issue if the beer does present a flaw and, hopefully, also deduce whether the problem needs to be addressed at the brewery or within the supply chain. Lastly, soliciting the beer from the customer at your own expense demonstrates that you are willing to go the extra mile to ensure the quality of your beer, which sends a positive message to the consumers supporting your brewery.

Upon receiving beer back from a customer, first take into account the nature of the complaint. If the consumer complained that the beer made them ill, do not put it in front of your panelists! Find other ways to assess it, such as analytical testing methods, rather than potentially serving something harmful to your panelists. If the complaint is regarding a more benign issue, such as simply tasting off, you can serve the beer to your panelists, though you should still evaluate the sample when preparing to serve it to ensure that it is safe for consumption. Assuming that the beer does not present any safety issues, perform a description test with your panel to try and ascertain the issue. Many consumer complaints stem from old or mishandled beer, though consumers may also catch a slow-developing microbial issue that may not have been apparent when the beer was first packaged. The results of a description test will hopefully give you enough information to both respond to the customer and determine if further steps need to be taken at the brewery.

The second tier of beer library maintenance involves only retaining beer from certain batches, so it does not require too much space in total. Specifically, you want to keep any batches flagged by a significant portion of panelists when initially run through product release TTT testing. Depending on the amount of space you have, you may only be able to hold on to batches that outright failed a product release panel but were still released for sale. If you have a bit more space, you can lower the threshold of not TTT responses that would trigger you to hold onto samples from a given batch. Over time, you may find that you are able to correlate certain levels of not TTT responses to customer complaints, which will allow you to better determine which batches to hold on to. With any problematic batches that you retain, you should monitor the beer periodically using the shelf life TTT test described later in this chapter. Depending on the initial issue and what the brewery decided to do with the batch, this additional monitoring may trigger a response if the issue worsens over time. For example, if your panel noted a mild diacetyl issue in a packaged beer and the brewery decided to keep the beer in-house to only sell through the taproom, an increase in diacetyl at a one-month shelf life test might result in pulling that beer from service entirely.

The third tier requires significantly more storage capacity but it offers you the greatest ability to respond to issues if and when they arise. At the third tier, you should keep a six-pack of beer from each batch. At a minimum, you should hold on to the beer for the stated shelf life of the brand, though, if possible, hold on to the beer for an entire year.

One common question when building a beer library of any size is whether to store the beer warm (i.e., at ambient temperature) or cold (i.e., refrigerated). In a perfect world, you would be able to set aside enough beer from each batch to store some warm and some cold. In most cases, space constraints will require you to choose one or the other, or the layout of your facility may dictate whether you are able to maintain a refrigerated beer library or have to store your beer at ambient temperature. However, if you have the choice, you should weigh the pros and cons of each approach and choose the storage conditions that make the most sense for your beer library.

If you elect to store your beer warm, your beer library will represent a worst-case scenario view of your beer and in many cases will probably taste worse than beer out in the trade. Oxidation flavors will develop more rapidly. If the beer contains any microbial spoilers, they will grow and alter the flavor of the beer more rapidly than in beer stored under refrigeration. If you get a consumer complaint on a batch, testing warm-stored beer will make it easier to find any issues present, as they will have developed to a greater extent than in refrigerated beer.

The primary drawback to maintaining a beer library at ambient temperature is that it does not do a very good job of approximating the beer that the customer receives in most cases. A number of factors can impact the quality of your beer in the field, from conditions at distributors' warehouses to the amount of refrigeration available at retail outlets selling your beer. All that aside, cold storage will, all other things being equal, better emulate the conditions your beer experiences prior to consumption.

Some breweries try to take a hybrid approach by storing the beer warm for a month (or some other period of time) before moving the beer into cold storage in an attempt to mimic the conditions that the beer sees while out in the trade. However, it is rather tedious and challenging to manage your beer library in this manner. What is more, it also tends to skew any data relating to those beers because it will look like the quality of the beer degrades rapidly for a month and then stabilizes, when in reality the change in rate of flavor degradation is due entirely to storage temperature effects. In short, I do not encourage the use of a hybrid storage approach.

When assembling a beer library, pick one set of conditions and stick with it for all beers. This way, you can compare any data collected from one batch against other batches, knowing that both experienced the same set of conditions throughout their respective shelf lives.

DETERMINING SHELF LIFE

For most consumer products, shelf life is defined by the length of time the product will remain fit for consumption and retain its desirable sensory properties, typically with the focus of attention being safety (Rogers 2010, 144). However, fitness for consumption is not usually a concern with beer. Due to a number of intrinsic characteristics, namely ethanol content, low pH, and the presence of hop iso-alpha acids, beer is naturally protected from many potential spoilers, and no known human pathogens (microorganisms that make us sick) can grow in beer (Vriesekoop et al. 2012, 335). Consequently, changes in a beer's sensory qualities typically determine its shelf life. However, designing experiments in an attempt to define a beer's shelf life presents some difficulties, because the final determination of a beer's shelf life is largely a subjective decision.

Beer changes as it ages, in some cases quite rapidly. Though the timeframe will vary from brand to brand, a well-trained panel can often pick up on differences in a batch as young as two weeks beyond packaging. As such, the question is not whether the beer tastes the same at the end of its stated shelf life as it did when it came off of the packaging line—it does not. Rather, you want to determine how long the beer remains close enough to the brand profile that you are still comfortable selling it to customers. What "close enough" means is open to interpretation, and there is no one correct answer when it comes to setting the shelf life for a brand. Similarly, no universally agreed upon method exists for performing shelf life assessment of beer. However, by modifying some of our existing tests, we can at least produce data that can shape the discussion around how long a given brand should remain in the market. The decision of when to stop selling a brand carries financial implications and should be a decision informed by the sensory program and made by management.

Approach shelf life testing with clear objectives in mind. Any assessment of shelf life you ask your panelists to perform should be part of a planned study for a given brand. The study might evaluate whether the brewery should increase or decrease the stated shelf life, or the study could attempt to determine whether a new treatment in the brewery had an impact on the shelf life of the beer. You should avoid conducting shelf life assessments at random without first thinking through how you will use the data produced. Some breweries will include shelf life samples in their standard panels, periodically tasting aged batches to see how they are holding up. Without an organized plan for sampling a brand across its shelf life, this approach typically does not yield actionable data and so is of little use to the brewery. Any time you plan on running a new sensory test you should consider whether the data produced from the test will actually inform decisions made by the brewery. If not, then you probably should not perform that test. You should also periodically revisit those tests you consistently use to ensure they still provide you with useful, actionable data.

IS THERE ANY VALUE IN FORCE-AGING A BEER?

In an effort to expedite the testing process, some breweries will perform shelf life tests on "force-aged" beers, that is, beers that have been oxidized rapidly through high temperature storage (often in excess of 32°C, or 90°F) in an attempt to mimic the flavors produced by long aging. While the rate of oxidation does indeed increase with increasing temperature, the specific mix of oxidation-related flavor compounds varies depending on the storage temperature (Vanderhaegen et al. 2006, 358). More importantly, some of the compounds that form during high-temperature storage may never appear in beer stored under refrigeration. In short, force-aging beer at high temperatures does not accurately reflect the way that beer changes during storage under typical conditions—it only reflects the way beer ages when stored at extreme temperatures. I recommend against using force-aging to assess the shelf life of your beers.

Shelf Life Sensory Testing Challenges

While sensory texts do not present a universally accepted test for assessing the shelf life of a product like beer, there are several different methods currently practiced in the trade. Some breweries will simply run aged beer through their standard product release TTT panel. The rationale for this method is that as the beer ages and changes a higher and higher percentage of panelists will mark the beer as not TTT. Mapping data in this way will show you how each brand changes over time, which can inform the decision as to how long the beer should remain in

the trade. However, experienced panelists become quite sensitive to slight differences in each brand. If aged samples appear in your product release panels, panelists will assess them as if they were fresh off the packaging line, which neither accurately reflects the condition of the beer nor the type of judgment you want to inform. Panelists will begin to mark the beer as not TTT as soon as the flavor profile starts to change, often within the first month after packaging. Consequently, your results will show that the aged sample is not true-to-target within a relatively short period of time, which will not help the brewery very much in making a decision.

Additionally, this approach can potentially impact your panelists' performance on their standard product release TTT tests. If you routinely add aged beer to regular TTT tests, over time panelists will subconsciously begin to expand their concept of the acceptable range of variation within that brand, even if they flag the aged samples as not TTT. This will make them less sensitive to issues in fresh beer, which can compromise the integrity of your panel's quality control function. Since quality control and product release monitoring should be the paramount goal of your sensory program, you should avoid serving aged samples on your product release TTT panels.

CAN DIFFERENCE TESTING BE USED TO ASSESS SHELF LIFE?

Based on its name alone, difference testing may initially seem like a good candidate for testing how beer changes over time. However, difference tests are far too sensitive to give us valuable information regarding beer shelf life. Difference tests detect minute, barely perceptible differences between samples. As a beer ages, its flavor changes rapidly. In a difference test, panelists will likely have no trouble distinguishing fresh samples from samples aged for much longer than a couple of weeks, meaning you will find a highly significant (and largely meaningless) difference between samples long before you would consider pulling the beer from shelves. Beer shelf life testing must accommodate the fact that a beer will change over the course of its usable shelf life, which means difference testing is not well-suited to the task.

The difficulties encountered with the standard TTT test get at the heart of what makes shelf life testing such a challenge. You already know that the flavor of the aged samples will differ from the fresh beer—that is not the question you are trying to answer. Ultimately, you want to know how quickly differences appear and how different is too different. The type of tests that only require panelists to give binary yes/no or stop/go answers will not give you enough information to answer either of these questions. Instead, you should employ some type of descriptive testing. A good shelf life test will tell you how a beer changes over time and how quickly those changes occur. You can then present that data to management, allowing them to determine when the brand has drifted too far from its original flavor profile.

One possible approach would be to periodically run description tests on a given batch to map exactly how the beer changes over time. For example, you might perform a description test each month for six months (or nine months, or however long you anticipate the beer to remain saleable). Equipped with these descriptions, the shelf life discussion becomes more straightforward. By comparing the descriptions of the aged beer against the profile of the brand when fresh, management can make an informed decision regarding the point at which the brand has changed too much to remain in the market. However, I would not recommend this approach for most breweries. While this method provides useful data, description tests require a significant amount of panelists' time because you can only perform a few of these tests within a single session. If you have to perform periodic description tests across all of the brands you carry, you may find yourself trying to work through more tests than your panel can realistically handle. Using the description test to map shelf life may work for some breweries, but in many cases it is too tedious to be feasible.

Shelf Life True-to-Target Test

The most effective method I have encountered is a modified TTT test, which I will refer to as the shelf life TTT test (as opposed to the product release TTT tests you use to assess fresh beer before it leaves the brewery). The shelf life TTT test combines elements of a standard TTT test and a description test to provide a significant amount of actionable data while inducing less cognitive fatigue in panelists than a full description test.

The setup for a shelf life TTT test mirrors the standard product release TTT test. Like a product release test, you should present samples with the brands identified and give panelists access to the brand target descriptions, but withhold all other identifying information (e.g., batch code, packaging type) as you would normally. However, you should explicitly state that this is a shelf life TTT test, which alters both the way that panelists assess the beer and the responses that they give. Panelists should assess each modality, comparing their perceptions against the target description of the brand when fresh. Ask panelists to provide lots of feedback on each modality, noting in particular whether the attributes initially present have increased, decreased, or remained the same; in addition, panelists should note any new flavors or characteristics that have emerged. By soliciting this kind of feedback from panelists, you will receive a level of detail similar to what you might obtain from using a description test. However, allowing panelists to reference the fresh brand target descriptions to guide their assessment will reduce the overall cognitive strain of the test.

Similar to the product release TTT test, panelists should use the overall description of the brand to make their final decision of whether the aged sample remains true-to-target. The "overall" category lists the flavors that must be present for a brand to stay within specifications—loss of these key characteristics might indicate the point at which the brewery should pull the brand from shelves. While panelists should tolerate a bit more variation in their evaluation of aged samples, they should use the presence or absence of these key traits to guide their final assessment. An example of a completed shelf life TTT test ballot is shown in figure 11.1.

Due to the nature of a shelf life test, the test itself will naturally bias panelists by encouraging them to seek and find aged flavors, whether or not there are any actually present. To combat this bias, inform panelists during training on this test that you will present them with beers across a wide range of different ages, potentially even including fresh beer. When you begin administering shelf life TTT tests, follow through on that statement. Most panelists hate being wrong, and the knowledge that you might include a non-aged sample will keep them on their toes. By occasionally using a fresh sample in a shelf life TTT test, you will reduce the impact of expectation errors and selective perception biases on your data.

Sample 395 is a shelf life sample of American Wheat Beer. Assess each modality against the target description for a fresh sample of American Wheat Beer. For each modality, please provide feedback on ways in which the beer has changed, specifically commenting on whether the original traits have increased, decreased, or remained the same and noting any new flavors or characteristics that have emerged. After entering your feedback, determine whether that modality remains true to brand. After assessing each modality, make a final assessment of whether the beer overall is still true to brand.

Target Appearance: Straw color, low haze, white foam

☐ TTT ☑ Not TTT

Feedback: *Dark gold color and significant haze present (beer is nearly opaque).*

Target Aroma: White bread and water cracker with light pear and banana

☐ TTT ☑ Not TTT

Feedback: *Malt flavor diminished, banana esters increased, development of honey-like oxidation flavors overshadows other notes.*

Target Taste: Low sweetness, moderately low bitterness, no sourness

☑ TTT ☐ Not TTT

Feedback: *No change.*

Target Mouthfeel: Medium body, moderately high carbonation

☑ TTT ☐ Not TTT

Feedback: *Carbonation slightly lower than usual.*

Overall

Shelf Life Target: Beer should be free from defects and must have bready malt flavor and fruity ester character.

☐ TTT ☑ Not TTT

Feedback: *Key flavors still present though out of balance, significant oxidation flavors present.*

Figure 11.1. Example of filled in shelf life TTT test ballot.

Just like the product release TTT test, you should remind panelists during training that they are not making a determination of whether or not to sell a given beer when voting TTT or not TTT. That decision lies with management; panelists should keep this in mind to allow them to remain as objective as possible.

Compared to a product release TTT test, the final determination of a shelf life TTT test involves a bit more subjectivity, as it gets at the question of how different is too different. Putting this subjective decision into the hands of your panelists can potentially lead to inconsistent results, as each panelist may have a different idea of when a brand has strayed too far from its target. As a consequence, p-charts of shelf life TTT test data offer less value compared to those derived from product release panels. You will likely find that the descriptive data gleaned from this test serves you better than the TTT/not TTT responses.

However, the shelf life TTT test also conforms to one of our original guiding principles for our sensory program—whenever possible, aim to use tests that are easy and quick to execute and yet accurate at finding defects. Once you have found a defect, further testing can be performed. A high percentage of not TTT responses in a shelf life TTT test does not necessarily indicate that the beer is beyond its useful shelf life. Rather, it signals that you should perform further testing, perhaps conducting a full description test. With data in hand from an assortment of tests, the brewery can confidently make a well-informed decision as to how long each brand should remain in the market.

12
APPLYING SENSORY TO RECIPES AND INGREDIENTS

As your sensory program matures, you can expand your panel's operations beyond solely monitoring beer to address questions relating to your raw materials. How can we ensure our hops are fresh before we use them? How do we describe the flavors produced by our base malt? Which variety of coffee should we use in our new coffee porter? To answer these sorts of questions, we will leverage familiar tests like the description test, extend the true-to-target (TTT) methodology, and add several new tests and techniques into our toolkit, each one unique to a specific ingredient or raw material. Some of these tests will assess the quality of a batch of an ingredient, helping you avoid purchasing substandard ingredients or using them in your beers. But this chapter also introduces applications of sensory testing for a purpose other than beer quality control. Rather than directly assessing fitness for use, several of the raw material sensory tests we shall explore here get at the characteristics of the ingredients themselves. These tests will give you a better understanding of the sensory properties of your ingredients, which in turn will help the brewery make better beer.

Suppliers of raw materials like malt or hops will typically provide a specification sheet for each batch. A specification sheet provides a wealth of quantitative information—for example, a malt specification sheet for a given batch will normally give you the color, moisture percentage, diastatic power, and soluble nitrogen ratio, among other things. However, most specification sheets do not provide much in the way of sensory information or flavor descriptors. A quick survey of different maltsters shows most malts described as a combination of "clean," "sweet," and some level of "malty," which does not offer much to a brewer trying to compare two different types of malt. Compare this to the wealth of descriptors found on DraughtLab's Base Malt and Specialty Malt Flavor maps (over 200 terms!) and it quickly becomes clear that we can do better than this.

Hop vendors typically do a better job of providing specific flavor descriptors for different hop varieties, but they only apply these descriptors at the varietal level (e.g., a set of flavor descriptors used to describe Cascade hops, a set for Mosaic, etc.) While better than nothing, it does not come close to capturing the variation in hop flavor and aroma you might find within a given variety from year to year or even from farm to farm. Brewers visit hop producers during harvest each year to select specific lots for a reason. Sensory characteristics can vary significantly between different lots of the same variety, so simply relying on the flavor descriptors provided by a hop vendor will not yield optimal results.

Furthermore, even if a raw material supplier were to perform rigorous sensory analysis of their ingredients and supplied you with a specific description from their panel, their choice of vocabulary may not match up exactly with your panel's lexicon. Performing sensory tests on your raw materials offers the only way to guarantee that you and your production team truly understand the flavors that each ingredient imparts. Your brewers can use this information to develop new recipes and to tweak existing recipes, for example, if they wanted or needed to replace the base malt in one of your beers. The test methods differ greatly from ingredient to ingredient, in many cases attempting to replicate the way that we use each ingredient in the brewing process. When you begin to introduce sensory testing for your ingredients, do not worry about trying to do everything all at once. Incorporate some of the simpler checks into your process at first (e.g., go/no-go for water and hops), then start to work through the ingredients most important to your beers.

RAW MATERIAL SENSORY TESTING

Malt

In small breweries, routine testing of malt quality can be accomplished within the brewhouse without needing to regularly send malt samples to sensory panelists for formal evaluation. Encourage your brewers to give malt a quick visual evaluation and to taste a small amount prior to brewing with it. You should not need to go to the length of creating target profiles for each malt that you use; your brewers will naturally come to know what to expect from each type of malt. You also want to avoid the brewers going overboard here—if you are using 55-pound bags of malt, your brewers do not need to make sure they taste from each and every bag. Occasional sampling while milling in should suffice. This sort of informal inspection might not seem like much, but it can catch a batch-ruining defect, such as malt that has gone sour as a result of bacterial contamination.

If you need to formally evaluate the sensory characteristics of a given type of malt, you can use the hot steep method from "Sensory Analysis 14" in *ASBC Methods of Analysis*.[1] Introduced in 2017, this method replicates the conditions of the mash on a small scale to produce wort for analysis. While the method does not account for changes that occur during boiling or fermentation, the hot steep method can give you a reasonable approximation of the flavors that a given malt will impart to beer.

The hot steep method begins with preparation of a standardized wort by mixing ground malt with hot water to perform a brief saccharification rest. To account for the increased flavor intensity and reduced diastatic power of specialty malts, the method recommends steeping specialty malts along with a portion of base malt. For crystal malts or toasted malts, such as biscuit malt or Victory malt, you should use a 50-50 mixture of base malt and specialty malt; heavily roasted malts, such as chocolate malt or black patent malt, should make up a maximum of 15% of the grist. Following the short saccharification rest, strain the mash through filter paper to produce a wort for tasting. (Refer to *ASBC Methods of Analysis* "Sensory Analysis 14" for a full description.)

Following preparation of the wort, serve it to your panelists and ask them to perform a description test on the sample. However, instead of using your normal beer lexicon, you should use a lexicon that more narrowly focuses on the types of flavors found in malt. I recommend using the malt flavor maps produced by DraughtLab, but you can build a custom malt-focused lexicon with your panelists if you prefer. When performing sensory analysis of the wort, focus your panelists' attention primarily on aroma description, as this best captures the flavor attributes malt imparts to beer. However, the other modalities can offer useful information as well. Appearance can give an approximation of the color contribution you can expect from a malt. The wort should generally taste sweet, and the presence of other tastes like sourness or bitterness may signify an issue with that batch of malt. Important mouthfeel parameters include body and astringency. Higher body and viscosity typically indicate a less modified malt, while astringency may result from issues with the malting process and can negatively impact your beers.

While the hot steep method can be used to assess the characteristics of a single malt in isolation, it works especially well for comparing different malts against one another. Some of the panel leaders I spoke to discussed using the hot steep method to choose specialty malts for a new recipe, compare similar products from different maltsters, and even replace the base malt of an existing flagship beer.

---

[1]   *ASBC Methods of Analysis* (online), "Sensory Analysis 14. Hot Steep Malt Sensory Evaluation Method," approved 2017 (St. Paul, MN: American Society of Brewing Chemists), doi: 10.1094/ASBCMOA-Sensory Analysis-14 (subscription required).

Hops

Unlike malt, the current methods in the literature for evaluating hop flavor and aroma primarily serve to identify defects and do not accurately convey how the flavors from a specific batch of hops will manifest in a beer. This primarily stems from the fact that the brewing process—particularly the application of heat and the action of fermentation—change the flavor compounds found in raw hops. During hop selection, brewers typically utilize the hop rub technique, vigorously rubbing hop cones in between their palms to break open the lupulin glands and volatilize the hop essential oils with the warmth of their hands. While an experienced brewer can use this method to select a crop of hops that they like, the hop rub does not serve us well in a sensory panel setting, not even for the purpose of detecting defects. The method lacks precision, in that the degree to which the hops are pulverized varies from person to person. Also, anything contaminating your hands—such as soap, hand sanitizer, lotion, or residue from testing a previous batch of hops—can affect the flavors you perceive from the rub. Fortunately, the ASBC offers a similar yet more exacting method for assessing the aroma of raw hops, the "Hop Grind Sensory Evaluation Method."

The hop grind method uses a blade grinder to pulverize a small sample of hop cones or pellets, which are then placed in a sample jar to allow for evaluation.[2] This method has a number of advantages over the standard hop rub. It uses less hop material, it creates significantly less mess, and it produces a standardized sample that panelists can evaluate against other samples without the risk of aroma carryover.

The hop grind method can function as a quick quality check on hops prior to their use in beer. As hops age, they may begin to develop isovaleric acid, which, if present, is readily exposed by the hop grind. Some breweries will even keep a small grinder on the brew deck to perform a quick hop grind prior to adding hops to the kettle. This quick test has the potential to prevent a bad bag of hops from ruining an entire batch of beer. Further, this brief task has the effect of focusing the brewer's attention on the task at hand. I have heard stories from multiple breweries that had to dump or adjust batches of beer as a result of a brewer adding the wrong variety of hops. Adding the hop grind method to your standard brewing procedure can greatly reduce the likelihood of such a mistake.

The ASBC Methods of Analysis also includes the "Hop Tea Sensory Method" for assessing hop aroma qualities.[3] This method produces a hop tea by steeping ground hops in room temperature water in an attempt to better approximate the aromas produced by dry hopping. However, a simple tea of hops in water fails to adequately simulate the flavors and aromas that hops contribute to beer, missing the interactions between yeast and hops, hop aromas and malt flavor compounds, and hop aroma and bitterness, not to mention it lacks the impact of carbonation.

While certainly a bit more involved, the best method for assessing dry hop aroma involves actually dry hopping beer on a small scale. I present this method with permission from Bennett Thompson at Half Acre Beer Company, where he uses this technique to assess new batches of hops. First, choose a beer from your portfolio to serve as the canvas for your dry hop. Ideally, the beer should be low in hop aroma and should not provide a lot of competing flavors from malt or yeast—something like an American blonde ale, a Kölsch, or a German-style Pilsner would work well. Begin by pulling about one gallon (3.79 L) of in-process beer from a fermentor once primary fermentation is complete. Transfer the beer to a small carboy and dry hop at a rate of about two pounds per US barrel, equivalent to about one ounce per gallon (7–8 g/L). Observe proper practices for limiting oxidation, including careful pouring and racking, and purging the headspace of the carboy with carbon dioxide. Cap the carboy with an airlock and allow it to sit at ambient temperature for three to five days. Make sure to give the beer adequate time to condition to allow the yeast to clean up any diacetyl or acetaldehyde formed during this step. Once the dry hop is complete, transfer the carboy to cold storage to crash the beer, then rack the beer into a sanitized two-liter bottle. Using a carbonation cap, carbonate the beer in the bottle and serve it to panelists for evaluation using the description test.[4] This method requires

---

2   *ASBC Methods of Analysis* (online), "Sensory Analysis 16. Hop Grind Sensory Evaluation Method," approved 2018 (St. Paul, MN: American Society of Brewing Chemists), doi: 10.1094/ASBCMOA-Sensory Analysis-16 (subscription required).

3   *ASBC Methods of Analysis* (online), "Sensory Analysis 15. Hop Tea Sensory Method," approved 2016 (St. Paul, MN: American Society of Brewing Chemists), doi: 10.1094/ASBCMOA-Sensory Analysis-15 (subscription required).

4   B. Thompson, interview with author, Chicago, August 9, 2019.

significantly more work and forethought than simply preparing a hop tea, but it is the only method I know of that allows you to accurately assess the flavors and aromas of a given lot of hops.

Yeast

When it comes to yeast, we assess most parameters, such as yeast viability, vitality, and cell count, using laboratory techniques. However, we can also use sensory methods to evaluate the quality of a yeast culture. As with malt assessment, you can set up a quick spot check your brewers should use while handling the culture, specifically while harvesting yeast from a given batch. In this spot check, your brewers will assess both the appearance of the culture as well as its flavor. When harvesting yeast, you want to make sure that your brewers are pulling healthy yeast while discarding trub and dead yeast cells. Healthy yeast will be creamy and beige in color, while trub and dead yeast usually range from green to dark brown.

Brewers can also evaluate the flavor when they are harvesting the yeast to make sure nothing is off. To taste the yeast, remove a small amount of yeast from the culture using aseptic sampling techniques. Dip your finger or a clean spoon into the sample and taste the yeast, but avoid consuming too large of a sample due to the laxative properties of fresh yeast. Fresh yeast should be bright and spritzy with no off notes. Meaty flavors or umami taste can indicate autolysis and dead yeast cells, while sourness or phenolic notes like 4-ethylphenol (4EP) can result from bacterial or wild yeast contamination.[5] As with the malt spot check, you do not need to develop target profiles for your yeast; over time, your brewers will become familiar with the way your yeast tastes when healthy.

Unfortunately, the flavor of the raw yeast itself does not really correlate to the flavors it will produce during fermentation. Due to the impact of fermentor geometry on yeast behavior during fermentation, the flavors produced by a given yeast strain can vary dramatically between a small test-batch fermentation and a full-size batch of beer. As a result, it is hard to build an understanding of the flavors and characteristics of a yeast strain until you are already brewing with it. Whether you are trying to select a new yeast strain to produce a new brand or trying to replace an existing strain, you will often have to lean on your yeast supplier(s) at first to tell you the expected flavors and characteristics. Once you have begun using the new strain in the brewery, you can perform a description test with your panel to better understand its flavors.

Water

While you will not use sensory evaluation to understand the general characteristics of your water, you should use sensory to test the quality of your water every single day. Some breweries will put water samples in front of a sensory panel, but your testing does not need to be quite that formal. Instead, your brewers should perform a simple go/no-go test prior to using water in the brewery for the first time each day. At least two people should taste the water, looking for common water-derived off-flavors. Chlorine/chloramine, metallic, and earthy flavors are the most likely issues your tasters will encounter. The tasters should also assess the appearance to make sure the water is both clear and colorless. If you have multiple sources of water in your brewery (e.g., your hot liquor tank gets water from a different source than your packaging jetter), test each source prior to using it. Given that water makes up the majority of beer by weight, any issues with your source water will generally ruin a batch of beer. This test is quick and incredibly simple—make it part of your brewers' standard operating procedure for beginning a shift.

Carbon Dioxide

While not usually discussed alongside the four main ingredients, carbon dioxide (CO_2) also plays a role in each and every one of your beers. If you force carbonate your beers, you should treat your CO_2 as you would any other ingredient. Your CO_2 supply can become contaminated with a wide variety of other gases, including acetaldehyde, diacetyl, and a range of sulfury and earthy off-aromas. Since these aroma compounds are odor active at such low levels, it does not take much to spoil your entire gas supply, and a contaminated gas supply will readily impart its contaminants to your beer. To assess your CO_2 supply for contamination, you can bubble gas through water and then present that water to panelists for evaluation. The simplest procedure involves filling a bottle or growler

---

[5]   Erik Fowler (White Labs), email correspondence with author, September 25, 2020.

two-thirds of the way with distilled or bottled water and bubbling CO_2 gas through the water for 5–15 minutes. Cap and cool the carbonated water in the refrigerator and then serve samples of the water to panelists, asking them to smell and taste the water for the presence of any off notes.[6] Many large breweries will test their gas source using a sensory panel every day. I recommend testing periodically, perhaps quarterly or monthly to start. It is an easy test to prepare and execute, and it can save you a lot of trouble if you end up with a contaminated CO_2 supply.

Other Raw Materials

Brewers today work with an ever-expanding palette of novel ingredients in addition to malt, hops, yeast, and water. Although once primarily confined to small, eclectic breweries, ingredients like fruit, spices, chocolate, coffee, herbs, and other adjuncts now find use in beers produced by breweries of all sizes. Just as with other ingredients, we can use sensory evaluation to better understand the impact that these foodstuffs will produce in a beer, to select among several varieties or sources of the same ingredient, or to evaluate the quality of a batch of an ingredient. Per usual, the specific question you are trying to answer will determine the appropriate sensory test to use.

If attempting to develop a new beer with a new raw material, you may want to test a number of different suppliers or varieties of the ingredient to determine which one you like the best. For example, imagine you are developing a new coffee stout and need to select a source for the coffee. In this case, you can brew a small amount of beans from each of several samples and use a hedonic test, which allows you to compare panelist liking scores across several different coffee varieties. On the other hand, you may want to highlight a specific characteristic of your unique ingredient. Perhaps you want to select a coffee varietal that expresses red berry aromas. In this case, you can use either attribute scaling or a ranking test. In attribute scaling, you ask panelists to assess each coffee and to specifically rate the level of red berry aroma (using as many scale points as you feel is appropriate). If you desire, you can follow this up with a hedonic test to collect liking information as well. In a ranking test, you ask panelists to rank the coffee

samples from highest to lowest level of red berry aroma. Either of these two tests will help you identify the coffee with the highest level of red berry aroma, though panelists might struggle with the ranking test if working with more than five or six samples.

In some cases, you may want to test the brewery's supply of a novel ingredient to check for flavor degradation, especially with ingredients like spices and extracts that can decline significantly with age. To assess the extent of degradation, we turn to yet another novel application of a familiar test—the TTT test. Just like with your beer brands, you first create a target profile with your panel that covers the characteristics of a raw material, then you test against that profile to determine whether or not significant degradation has occurred.

As with the standard TTT test, you first need to create a profile using the description test. Obviously, you will not be tracking the same characteristics as you would for a beer. In most cases mouthfeel—and perhaps even taste and appearance—will be irrelevant. However, if any of these modalities have bearing on the attributes that define a quality sample of the ingredient, include them in your profile. You should also establish the overall characteristics that *must* be present in order for the ingredient to pass the test. For example, if you brew a witbier, experience may dictate that as soon as your raw coriander loses its lemon-orange citrus character it no longer produces acceptable beer. In this case, you can include in the overall TTT description that the coriander must exhibit lemon-orange aromas in order to be marked as TTT. With a target profile in place, you can run a raw material TTT test just like you would any other TTT test. Over time, you will learn how frequently or at what point you should test each ingredient for flavor degradation. Based on what you learn about the shelf life of your raw materials, you can also adjust the point at which you begin testing ingredients to eliminate unnecessary tests. For example, suppose you have never had an issue with coriander less than three months old—in that case, you might decide to begin testing the coriander at the three-month mark and continue to test it once a month until you either have gone through that batch or encounter an issue.

As with other beer ingredients, raw materials may require some amount of preparation before you serve them to panelists. Most fruit products, from whole

6   Bill Simpson and Cara Simpson, "Sensory evaluation of carbon dioxide in the brewery" (poster presentation, 2019 ASBC Meeting, New Orleans, LA, June 24–25, 2019).

fruit to purees to juices, can be served to panelists as is. As in the coffee stout example above, you can brew coffee beans to produce coffee for sampling. Grind spices for aroma sampling in a manner similar to the hop grind method (p. 137). Dilute flavor extracts or fruit concentrates in water or a sugar-water solution to allow panelists to smell and taste them at a reasonable intensity level. However you decide to prepare your raw materials, record the methods you use and stick to them to make sure your samples reflect similar preparation protocols from one evaluation to the next.

RECIPE ADJUSTMENT

From time to time, you may want or need to adjust a recipe. Perhaps you have been considering tweaking the hop bill on one of your IPAs to improve the flavor profile. Perhaps your current coriander supplier has gone out of business and you need to source a new variety to continue producing your witbier. You may want to reduce the cost of your flagship Pilsner by substituting your base malt for a cheaper alternative. Or maybe you are trying to scale up your tangerine pale ale and it is no longer feasible to juice and zest all of the citrus for the beer by hand. Whatever the reason for making an ingredient change, you should use sensory testing to guide the process.

These scenarios may seem like perfect candidates for using a difference test, such as a triangle test or a tetrad test. In truth, many large industrial food and beverage producers use difference tests to address these exact situations. However, as discussed in chapter 7 (pp. 88), if you want to pursue difference testing you must first validate that you can produce beer consistently enough from batch to batch that your panelists cannot differentiate between batches. For many small- and mid-sized breweries, batch-to-batch variation precludes them from using difference testing.

Assuming that you do fall among the small percentage of breweries capable of validating consistent production processes, difference testing *still* may not be the best choice for testing recipe adjustments. Difference tests are incredibly sensitive and can pick up on the smallest of differences. However, a difference test does not tell you whether the difference revealed actually matters. This leads us to a test that is better suited to evaluating recipe substitutions and one that we are already quite familiar with—our trusty product release TTT test! In running a prototype through your

product release panel, your panelists will compare the test batch against a target description rather than against another beer. Consequently, you do not need to worry about producing a control batch and a test batch at the same time, which is a tremendous benefit, especially if your production is on the smaller side and you do not usually produce multiple batches of a given brand at one time. Furthermore, since the product release TTT test already incorporates and accounts for some batch-to-batch variation, it will allow you to determine whether any differences resulting from your changes are large enough to matter.

Recipe adjustments can typically be grouped into one of two broad categories: changes designed to improve the beer, and changes designed with the goal of maintaining the beer's current profile. The TTT test can serve you in both of these scenarios, but your goal will affect the way you analyze the results. If you are attempting to make a change that improves the beer, your first objective should be that the prototype fails the TTT test. After all, if your panelists do not think the altered brand differs from its normal profile, you likely have not made a significant improvement to the beer.

If the test batch fails a product release TTT test, you can then use other tests to assess whether the changes made represent an improvement. For example, you can use a hedonic test on the existing beer and the test batch to compare overall liking scores for the two beers. If you have a taproom, you can even expand such a test beyond your panel to include consumer data sourced directly from your customers. Alternatively, if you are targeting a specific characteristic (e.g., you want to increase the tropical fruit aroma of a beer), you can use your panel to perform attribute scaling for that specific attribute or sample ranking based on that attribute. By comparing the test batch(es) to the existing beer, you can determine whether you have actually increased the specific characteristic you are targeting.

If you are trying to make an ingredient substitution without changing the flavor of the beer, your hope is that the prototype still passes a TTT test. If the prototype fails the TTT test, then it certainly would have shown a significant difference on a triangle or a tetrad, obviating the need to run a difference test. If the test batch passes the TTT test, you should still collect more data before moving forward with the change, especially if it is a significant change to a major brand. Brew a few additional test batches and continue running them through your

TTT product release panel. You never want to make a decision based on just one data point.

It is worth noting that even larger breweries with the capability to run valid difference tests will still run the test batches through their standard TTT panel, only submitting them for further difference testing if the beer first passes the TTT test. Remember, difference tests tell you whether or not a difference exists; TTT tests tell you whether that difference matters. If batch-to-batch variation prevents you from running difference tests on your beers, rest assured you can still reliably assess substitutions and adjustments using the TTT testing protocol.

PROCESS ADJUSTMENT

At smaller breweries, concerted efforts to tweak process parameters in an intentional way typically take a backseat to simply trying to produce consistent beer. However, as breweries grow, brewing efficiency becomes an increasingly important consideration. Once consistent brewing is achieved, brewers may begin tweaking various parameters in an attempt to improve efficiency, including but not limited to milling parameters, mashing parameters, fermentation parameters, and dilution rates. Similarly, in an attempt to reduce costs (or to expand into a new format), brewers may also experiment with different packaging materials. As with many recipe adjustments, breweries typically aim to maintain the current flavor profile of the brand whenever making a change. In order to proceed with a change that reduces costs or improves efficiency, breweries want to first ensure that the change does not negatively impact the sensory characteristics of the beer.

As with recipe adjustments, you can certainly use difference testing to assess a process change. However, I recommend once again beginning by running the test batch through your product release TTT panel. Using the TTT test will tell you whether or not the change pushes the beer outside of the range of acceptable variation for the brand. Assuming that the test batch passes, you can then determine whether or not you need to perform additional testing.

If attempting to test a process variation, remember to think like a scientist and only adjust one variable at a time. I have heard all too many anecdotes of breweries constantly adjusting various parameters and running tests on their beers at virtually all times, only to find their data completely unusable since they cannot determine which changes correlate to which effects. Additionally, perpetually tweaking your process will all but ensure a highly variable product. If you are always making changes, you should expect the consistency of your beers to suffer.

13
BUILDING
NEW BEERS

Many different factors can influence the decisions a brewery makes when developing a new brand. Small breweries, with fewer total decision makers, often use a relatively straightforward or unstructured brand development process. Perhaps your brewmaster wants to tackle a style that they have always wanted to brew, or maybe you leverage feedback from your taproom customers or sales figures from prior releases. However, as a brewery grows and the number of stakeholders increases, so too does the complexity of any new project. A new beer might still stem from a project that the production team wants to pursue, but pressure to design a new beer can also come from the marketing department deciding you need to brew a certain beer to fill a gap in your portfolio.

Regardless of who kicks off the development of a new brand, the basic objective is the same: produce a quality beer that meets a certain set of desired flavor characteristics. Though some exceptions exist, most of the beer brands on the market today are sold based on the merits of their flavor. And if a new brand's success relies on achieving specific flavor characteristics, then sensory testing will obviously help inform new product development. Your sensory program may have already helped facilitate this process through sensory profiling of raw materials. However, you can use your panelists in a number of additional ways to improve a new brand throughout the product development process.

DEFINING AND SHAPING A NEW BRAND

When building a brand, begin by designing a target flavor profile for the new beer. The target should specifically identify the desired key parameters of the beer. For example, if aiming to brew a New England IPA, you might determine that you want the beer to feature low sweetness and pronounced bitterness, with primary aromas of grapefruit and tropical fruit. Within the realm of aroma, it is better to stick with broader flavor terms during the development phase. Choosing descriptors like grapefruit and tropical fruit rather than more specific terms like "grapefruit pith" and "overripe pineapple flesh" will improve the likelihood that your brewers actually hit the target. Once the target has been agreed upon, your brewers should attempt to produce a test batch to match that description.

Development of a test batch is typically the first point in the product development process to involve your sensory program. At this stage, your panel can help identify whether or not your brewers managed to hit the target description. While you can attempt to answer this question in a few different ways, the most straightforward method uses a description test followed by a modified true-to-target (TTT) test. Begin by presenting the beer to your panelists blind and ask them to perform a standard description test on the beer. As always, avoid giving the panelists any information about the target profile or the beer itself to avoid biasing their responses. After the panelists have finished their descriptive profiles of the beer, you can then present them with the previously determined target profile and ask them whether the beer that they just described is consistent with the target description. In analyzing their responses, you should compare the composite panel description of the test batch against the target profile to give you some idea of whether the beer attained its goal. However, asking panelists to consider whether the target profile lies within the same realm as their blind description of the beer will give you another useful data point.

NEVER TOO SMALL

For small breweries—particularly those producing a lot of one-off beers—mapping out the creation of each new beer in this manner may seem too tedious to be worth it. However, the process of producing a target description does not have to get too involved. It can be as simple as asking your head brewer to write down the key characteristics they hope to achieve prior to brewing the batch. But setting up a product development process, however rudimentary, from the beginning will help immensely as you grow. Over time, more and more hands will get involved in the product development process. Establishing a system early on will help prevent disputes down the road by clearly identifying who makes decisions at each stage of the process. Additionally, and especially important for a small brewery, brewing to a written target profile will help your brewers improve. You will likely be able to sell the beer regardless of whether you hit your target exactly, but your brewers should still refer back to the initial goal once the beer is ready for release. In doing so, your brewers can assess whether or not they hit the mark and may be able to determine how to better meet their targets in the future. Without putting a target down in writing, the brewing team will likely drop their original intentions in light of the reality of the finished beer, ultimately robbing themselves of the opportunity to reflect and improve.

After analyzing panelist data, you should meet with the brewing team to determine next steps. Depending on panelist feedback, you may need to tweak the beer and brew another test batch. For example, going back to our New England IPA, if one of the desired flavors was grapefruit and none of your panelists noted any sort of citrus aroma then your brewing team missed the mark on that front and they need to figure out a way to build that flavor into the next batch. Alternatively, you may decide that you like the beer the way that it was made even though it did not meet the original target, in which case you can revise the target to match your panel's descriptive profile. This approach may work in smaller breweries or breweries that primarily sell their beer through their own taproom, but it becomes less viable as a brewery grows in size. At larger breweries, meeting a specific, predetermined flavor profile is often an essential component of brand development.

You can also use your sensory panel to assist product development by performing hedonic tests to compare different trial batches. Assuming you have managed to produce multiple beers that, while different, fall within the target profile, you can use a standard, nine-point hedonic test to determine which test batch your panel prefers. If you have a taproom, you can also run hedonic tests with your customers to see which iteration of the beer they like best. The benefit of testing new beers with your customers is twofold: in addition to determining which prototype beer your customers prefer, over time you may also be able to correlate consumer liking scores to certain flavors or characteristics, which will help with future product development.

DESIGNING YOUR PRODUCT DEVELOPMENT PROCESS

As a brewery grows in size, the product development process invariably becomes more complicated. Many small breweries follow the mantra that they brew the beers they like to drink. With scale comes salability and profitability considerations, which increases the importance of brewing to a specific, well-defined, and carefully designed target profile. As more people get involved in the product development process, the potential for disagreement and disputes increases exponentially. If you can establish a standard process for brand development early on, that process will naturally become integrated into the fabric of your brewery.

At larger breweries, it also becomes increasingly important to document the steps of the product development process to protect yourself in case a beer does not perform as well as expected once it reaches the market. Good documentation can serve to mitigate risk for both you and your program.

BUILDING A TARGET PROFILE

While a small brewery may rely on their brewmaster to determine what a new beer will taste like, decisions at larger breweries often incorporate input from management, brand managers, and the marketing department. Failure to consider the opinion of an important department can tank a brand before it ever leaves the brewery. In order to ensure success, you should solicit input from all key stakeholders to obtain buy-in for the new brand throughout the brewery. However, you also want to limit the window during which you take that input. If you are constantly receiving feedback from several departments, you may end up being pulled in several different directions, ultimately making it impossible to please anyone.

If you do not already have a standardized product development process in place, begin by identifying who in the organization will determine the attributes of the target profile for a new beer. Depending on the structure of your organization, this could be the owner, the management or leadership team, the marketing department, or the brewing staff. It may also vary from project to project. In order to succeed, every project should clearly identify who is responsible for initial development of the target flavor profile.

Once the initial target flavor profile is developed, you should solicit feedback from all key stakeholders of the brand development process. Regardless of who develops the target, the key stakeholders will likely consist of a static group of people. Assuming that is the case, you should maintain a list of these stakeholders to make sure you both solicit feedback and secure approval for the finalized target profile from each individual on the list. This period represents the window in which different departments can offer feedback to inform the target profile. Ideally, you should establish among the stakeholders the finite nature of this feedback period and document each department's approval at this stage of the process. This way, you can confidently begin brewing test batches to match the target without worrying that one of the key stakeholders will try to offer new feedback once the recipe has already been developed, tweaked, and finalized. Cultivate the expectation that stakeholders can and should provide input throughout the target development process but should refrain from offering additional feedback once they have signed off on a final target profile.

Once you have finalized the target, your brewers can set about producing the product and your sensory program can help affirm when your brewers have met the desired target. Once you feel comfortable that the beer matches your original goal, you should take it to key stakeholders for review. At this point, you are looking for sign-off rather than any additional feedback. Documentation can prove essential at this stage. Pointing to their prior sign-off on the target profile alongside panel data supporting that you hit the target should head off any further suggestions. Once key stakeholders have signed off on the beer itself, lock in the target description and work with the brewing team to scale up the beer so that it still matches that description.

Not every project will proceed smoothly along this path, and sometimes a key stakeholder may decide to insert themselves later in the process. However, by establishing a standard protocol for developing new beers you will significantly streamline the process and save yourself a lot of headache.

14
BLENDING WOOD-AGED BEER

In the world of beer production, blending can refer to a variety of practices. It can simply describe a brewer blending two turns of similar wort into a single fermentor. With slightly more purpose, that same brewer might blend two slightly different worts to mitigate a certain characteristic—for example, if the first turn came out more bitter than intended, the brewer can lower the bitterness of the second to achieve the ideal level of bitterness for the entire batch. Similarly, a brewer might blend two finished batches together to balance out a given characteristic or even to conceal a low-level off-flavor in one of the batches. These practices primarily come into play in larger breweries who regularly brew a set of core brands and have the ability to blend multiple batches together before one of them starts to get too old. This type of blending often employs analytical measurements of parameters like IBUs and percent ABV to achieve a consistent end product, in addition to sensory techniques, such as the true-to-target (TTT) test.

In this final chapter, I want to focus on how breweries use sensory techniques to blend wood-aged beer. Wood-aging of beer is an inherently variable process. Whether producing clean beer or so-called wild beer (i.e., fermented with organisms other than standard *Saccharomyces* strains), a single batch of beer split across several different vessels will develop into

several different beers. The blending process can be straightforward, such as blending all of the barrels from a single batch back together to minimize any differences that have developed. Or it can be overwhelmingly complex, for example, blending different batches, different vintages, or even entirely different beers together to produce something new. It all depends on the brewery. Regardless of the end product, the goal of the blender is to produce a finished beer better than the sum of the original parts, and blenders often lean heavily on sensory techniques to achieve this end.

That said, blenders play by a different set of rules than sensory scientists. In many ways, the process of blending wood-aged beer takes everything we have discussed about sensory work and flips it on its head. Sensory panels originally evolved to improve product consistency and prevent companies from having to rely on the palate or expertise of an individual to make flavor decisions. But at many breweries producing wood-aged beer, blending decisions are made based on the sensory evaluation of a single person, and rarely will you find more than two or three people involved. While this may seem contradictory, it is, at least in part, a direct consequence of the goals involved in blending these types of beer. We train sensory panelists to function as instruments, evaluating samples and providing data, not to make concrete production decisions. In many cases, blending wood-aged beers involves subjective, hedonic decisions, which can be challenging to make in a group setting.

However, just because we are leaving behind the confines of the sensory lab does not mean we should abandon the guidelines of proper sensory practice. While decision-making may come down to a single individual, many programs use multiple tasters to evaluate both individual barrels and blends of wood-aged beer. When appropriate, I will refer back to the guidelines, best practices, and advice given in the first chapter of this book, as much of it will still serve us well in this arena.

To prepare this chapter, I interviewed and consulted with blenders from a wide variety of different breweries in both the US and Belgium. While every blender I talked to has a slightly different blending philosophy, their practices and approach tend to coalesce around a few distinct schools of thought. When possible, I have synthesized information from several individuals to present broad recommendations, but in many cases I thought it best to let each blender's words speak for themselves. I have tried to pull together a variety of different ways that you can use sensory techniques to improve your ability to blend wood-aged beer, but ultimately you will have to put it into practice and learn from your own experience to achieve success.

SENDING BEER TO WOOD

As liquid moves through the brewing process, it gets more expensive each step of the way. Whether through additional ingredients added, additional packaging materials used, additional labor, or just the vessel space occupied, each step incurs a cost. This holds especially true when transferring beer to wood, as wooden vessels typically bear a hefty price tag. And, as any experienced producer of wood-aged beer will tell you, making barrel-aged beer often involves a significantly higher dump rate than standard beer production. Consequently, you want to make doubly sure that you do not send bad beer to oak in the first place.

In most cases, beer destined for wood-aging will undergo primary fermentation in stainless steel vessels before it ever sees oak. To avoid potentially sending flawed beer to barrels, each batch should at least undergo some rudimentary tasting prior to

transfer. For some breweries, this may entail a simple go/no-go check performed by a couple of brewers on the production floor to make sure that the beer tastes as it should, does not exhibit any off-flavors, and is, in fact, the beer they intend to send to oak. Other breweries elect to run a full gamut of quality checks, including sending the beer to the sensory department for formal evaluation and performing quantitative analytical tests, from gravity to pH to cell counts. At a minimum, you should perform a basic sensory evaluation akin to the bright tank spot checks used to send beer to the packaging line (see p. 121).

Production of spontaneously fermented beer stands out as an exception. Most breweries producing spontaneously fermented beer will, following inoculation in a coolship, transfer wort directly to oak for primary fermentation without first tasting the inoculated wort.

TREATING OAK AS AN INGREDIENT

Just as you do not want to send bad beer into good oak, you also do not want to send good beer into bad oak. When producing wood-aged beer, consider the wooden vessel itself as an ingredient. Like other raw materials, you can evaluate wooden vessels using sensory methods to ensure their fitness prior to use. Most brewers recommend evaluating barrels or other wooden vessels when they first get to the brewery. Upon arrival, you should perform both a visual inspection for mold, liquid, or debris, as well as an aroma inspection for mold, acetic acid, sulfur notes, or anything else unusual. That said, you should also perform a cursory inspection of each barrel prior to each fill. Jeremy Grinkey of The Bruery shared an adage picked up during his years spent working in the California wine industry: "Never put liquid into a barrel without first putting your nose in the barrel."[1]

However, sensory assessment of barrels can go beyond evaluating the vessels themselves. Marty Scott of Revolution Brewing described a unique approach to using sensory techniques to aid in barrel selection.[2] Revolution's sizable barrel program consists almost entirely of whiskey barrels, which only see one use each. Consequently, the brewery is constantly purchasing new barrels for aging beer. When tasting each barrel prior to blending, the brewers take notes on the wood flavor imparted by each barrel—does

1   J. Grinkey, telephone conversation with author, April 24, 2020.
2   M. Scott, telephone conversation with author, May 8, 2020.

it lean more toward a woody lumberyard or does it lend warm notes of coconut or vanilla? They then map this sensory data against the identities of each barrel to help determine which distillery's barrels give the most desirable set of flavors. While this process took years to refine, Revolution has now settled on a few specific types of barrels that provide the exact set of characteristics the brewery is after.

TASTING IN-PROCESS BEER

When beginning a barrel program, your impulse will likely be to taste everything all the time. At this point, each project is new and exciting, and you are unfamiliar with how the beer will change over time. However, as your program grows—both in age and in size—you will want to practice restraint and taste the beer less frequently. From a beer quality perspective, every time you sample from a barrel you increase the head space in the vessel and potentially introduce oxygen into the beer. And, just like with standard sensory work, you should not just taste beers without any sort of purpose in mind.

Early on, it is fine to taste your beers frequently. Following best practice, each tasting should attempt to answer some sort of question—learning how the flavors of your beers develop as they mature certainly qualifies as an important question to answer. However, every blender I spoke to noted that, over time, the frequency with which they tasted their stock diminished dramatically. Most experienced blenders will not taste a beer during the first couple months it spends in wood, and once they have a given stock or brand really dialed in, they may only begin tasting the beer when they think it is nearing maturity and ready for use. As an extreme example, Jason Perkins of Allagash noted that for Curieux, a bourbon barrel–aged tripel, nobody actually tastes the individual barrels prior to blending. The beer only requires a seven-week rest in oak to develop the amount of flavor that the brewers are looking for, and the large size of each batch mitigates any barrel-to-barrel variation. When first producing Curieux, Allagash would perform test blends for each batch, but at this point the process is so well established that test blends are not needed. Prior to

blending, the brewery lab team will evaluate each vessel for microbial contamination, and the brewers will also give the samples a quick sniff check for off-flavors. Assuming the beer passes these two tests, it gets added to the blending tank.[3] Ultimately, the frequency with which you taste your beer will depend on a variety of factors, from the amount of time you anticipate the beer will spend in oak, to the size of the vessel, to whether the beer is clean or wild. First learn how your beers develop over time and then let your experience serve as your guide.

Taking Notes

Different blenders vary widely in their approach to note-taking. If tasting barrels in the early stages of a beer's development, most blenders usually just want to make sure things are progressing as they should, so notes at this stage are often sparse. As vessels reach maturity and the time for blending approaches, blenders will often take more in-depth notes, though the type of notes taken and their level of detail depends largely on the structure of the blender's program and their specific style of blending.

At Avery Brewing Company, Andy Parker primarily blends large batches (sometimes hundreds of barrels) of well-defined brands. At this scale, he notes that you lose a lot of the barrel-to-barrel nuances that develop during maturation, so their process primarily involves tasting each barrel to determine whether it fits with the overall blend and kicking out any barrels that exhibit notable off-flavors. Consequently, note-taking is minimal, sometimes boiled down to the simple use of smiley faces, neutral faces, and frowny faces.[4]

Some brewers opt for a slightly more regimented system of note-taking. Averie Swanson of Keeping Together explains that she typically records topline flavors—perhaps the three or four most salient aromas that stood out from the beer—and then would note a variety of structural traits, such as the level of acidity, bitterness, oak character, and body.[5]

Beyond aroma and taste, several brewers stressed the importance of also monitoring appearance. Wood-aged sour beer typically develops turbidity during maturation due to the activity of various bacteria and yeast. When the beer begins to clear, that indicates

3   J. Perkins, telephone conversation with author, May 7, 2020.
4   A. Parker, telephone conversation with author, April 28, 2020.
5   A. Swanson, interview with author, Chicago, April 13, 2020.

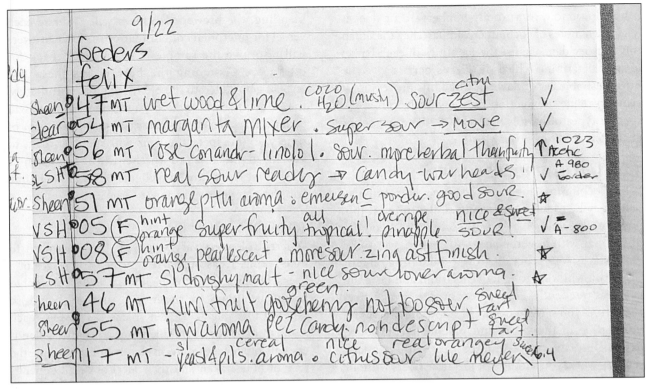

Figure 14.1. A sample of Lauren Limbach's tasting notes tracking the progress of several foeders. Note that in addition to recording key flavors, she also includes brief descriptions of each beer's appearance.

that the microbial fermentation is tapering off. Lauren Limbach of New Belgium Brewing states that she uses clarity as one of several signals that acidification is complete and the vessel has stabilized, noting that any wood-aged beer the brewery intends to package "live" (i.e., without pasteurization) should drop bright before it is transferred.[6]

I only encountered one blender—Frank Boon of Brouwerij Boon—who kept robust, thorough notes for each one of his *foeders*. Frank explains that he maintains a spreadsheet with 30 or so different attributes tracked for each foeder, covering things such as acidity level, bitterness level, and the presence or absence of a wide variety of different flavor compounds. Aside from a couple of analytically measured parameters like alcohol content and attenuation, each attribute is tracked through regular tasting of the foeders. With his decades of experience producing lambic, Frank says he can read his spreadsheet the way a musician might read the notes on a page, and can conjure a distinct flavor impression in his mind for any of his foeders just by looking at the sheet.[7]

Aside from Frank Boon, the blenders I surveyed noted that they started out taking more notes than they do today, gradually cutting down the initial volume of notes taken once they found that certain things did not serve them. Andy Parker mentioned that for years he kept a spreadsheet with detailed tasting notes, gravities, and pH measurements for each of his barrels, but eventually came to find that none of that information actually helped him make better beer. His current bare-bones approach to note-taking is the result of him carefully considering exactly what information he needs to make the best beer possible at his blending program, coupled with years of experience.[8]

If you are just beginning a barrel program, I recommend against taking too much of a laid-back approach. When first learning to age and blend beer, you are better off taking too many notes rather than too few. Over time, you can scale back your note-taking as you learn what information you actually need.

6   L. Limbach, telephone conversation with author, April 14, 2020.
7   F. Boon, telephone conversation with author, June 5, 2020.
8   Parker, telephone conversation.

Using Multiple Tasters

While I do not recommend running barrel samples through your sensory panel, using multiple tasters to assess beer both during development and blending offers some of the same advantages as panel tasting. No taster possesses a perfect palate; they each have their own strengths, weaknesses, and blind spots. While a number of excellent blending programs rely entirely on a single individual, just as many use a small group of tasters to assess their beers and shape their blends. Since tasting in-process beer is not as high stakes an activity as selecting beer for a finished blend, most breweries only have one person taste samples on an informal basis as the beers mature. However, many programs will shift to a more formal setting with multiple tasters when it comes time to assemble a blend.

To build consistency into your program, it is best to select a few tasters who always participate in tastings that involve wood-aged beer. In most cases, your team will consist of production staff members who oversee the barrel program. With a small group of tasters working together on a regular basis like this, a shared flavor lexicon will naturally develop, much as it does within a group of sensory panelists during training. Over time, this shared lexicon helps simplify communication and decision-making, ultimately leading to a more efficient blending process.

While tasting barrel-aged beer often occurs in a somewhat informal setting, implementing a bit of structure can improve the quality of each tasting session. At Jolly Pumpkin Artisan Ales, Ron Jeffries employs a two-part method for tasting with a group. First, each taster evaluates their samples privately, just like tasters on a sensory panel would. This helps to prevent mutual suggestion bias, which often comes into play when tasters evaluate beer while standing around a barrel together. Individual tasting leads directly into a group discussion, which helps the tasters consolidate their evaluations into an overall impression while also solidifying their shared lexicon. In addition to discussing how the flavors of each sample are developing, the discussion also allows Ron to share how he anticipates using each beer, helping members of the group better understand how he makes his blending decisions.[9] This produces more-informed, better tasters and serves to train the rest of the team to become better blenders as well.

---
9   R. Jeffries, telephone conversation with author, May 15, 2020.

Logistical Considerations

In an ordinary sensory panel, each logistical detail receives careful consideration, from the equipment used, to the tasting space itself, to the way the sensory data is managed and analyzed. Although tasting barrel-aged beer often proceeds without the same degree of structure, you should still consider these details to ensure optimal results. While some programs may appear to take a free-form approach to the tasting process, most blenders can articulate exactly why they do things in a certain way, indicative of a thoughtful, measured approach.

Tasting Space

In any scenario that requires focused tasting, you should endeavor to minimize sensory distractions as much as possible. Some brewers assessing wood-aged beer go to great lengths to label samples and serve them to tasters in a separate space to minimize competing aromas. But the reality is that most brewers taste beer in their barrel cellar, directly from the vessel. In most cases, it is simply not practical to try to transport samples to another location, especially given the large number of samples that blenders often have to evaluate. While an imperfect setting, you can mitigate some of issues associated with tasting in such an environment by using multiple tasters and consistently using the same set of multiple tasters. When it comes time to blend, most brewers will move out of the barrel cellar to perform the final tastings.

Number of Samples

Just like with standard sensory work, you should ideally limit the number of samples tasted at a time to reduce the potential for sensory fatigue. Unfortunately, in many cases, particularly in larger programs, the sheer volume of beer leads brewers to tackle large numbers of samples within a single session. The problem is compounded by many barrel-aged beers falling on the higher-intensity end of the flavor spectrum, whether due to the high alcohol content of a bourbon barrel–aged barleywine or the puckering acidity of a mixed-culture saison. If possible, try to keep the number of samples tasted in a given session to fewer than 10; some blenders impose such constraints on their tastings, but many of those I spoke to admitted that such limits are just not feasible typically.

If you find yourself needing to taste a lot of samples, use the following strategies to try to mitigate the effects of sensory fatigue. If possible, structure tastings around a given batch or brand of beer. By tasting similar products, you will find it easier to identify the nuances from one barrel to the next, ultimately reducing the amount of sensory fatigue that each sample incurs. Depending on the type of assessment you are performing you may not have to actually taste each sample, for example, when working based on aroma cues. Although this will still eventually lead to mental fatigue, it can help avoid the palate fatigue of tasting high-acid beers or the intoxicating effects of consuming high-alcohol beers.

Using multiple tasters can also improve the overall accuracy of a tasting session. While your tasters will still experience fatigue as they work through a large group of samples, having more than one palate assess each beer reduces the likelihood of a tired taster missing an off-note or a flaw in one of the beers. Harnessing multiple tasters also allows you to vary sample order from one taster to the next, mitigating potential sample order effects. Andrew Zinn of Wicked Weed shared that he sets up a panel with their lead blender, Jen Currier, by lining up all of the samples for a given session across a long bar. They then begin at opposite ends of the bar so that they taste samples in the opposite order to one another. In addition to limiting sample order effects, this also ensures that the samples Andrew tastes when most fatigued have been tasted by Jen at her freshest, and vice versa.[10] This prevents the final samples of the flight from getting short-changed in their evaluations.

Time of Day

Of the brewers I spoke with, many said that their busy schedules do not allow them to taste beers on any sort of schedule—instead, they fit in tastings wherever they have time. However, like with other types of sensory work, you should try to be consistent in the timing of your tastings if possible. This holds doubly true as the time for blending approaches, given that these later tastings carry more weight in shaping the final beer. Averie Swanson notes that she prefers to taste in the morning but typically ends up having to take notes on her beers in between other tasks. However, when it comes time to assemble a blend, she is almost ritualistic in her approach. She gets to the brewery early in the morning to assemble and taste her test blends before anyone else has arrived.[11] This sort of consistency helps ensure your palate is similarly calibrated each time you sit down to taste samples.

Recording Data

The size of your program and the type of notes you take will determine the exact infrastructure required to track tasting notes and other data. Methods vary widely among brewers, from digital methods, such as spreadsheets of varying levels of complexity or barrel management software programs like Barrel-IT, to more analog methods, such as using a notebook, printed cards on barrels, or simply chalk on barrels.

Within your sensory lab, tracking historical data is of the utmost importance, whether to help identify trends in the way a beer may drift over time or to respond to any consumer complaints or quality issues that may arise. However, due to the constantly evolving nature of wood-aged beer, tasting notes often do not provide much value aside from those taken immediately prior to assembling a blend. Harrison McCabe of Beachwood Blendery explained that, based on his experience, any tasting notes he takes typically only remain accurate for about two months. After that, the beer will have changed enough that the old tasting notes no longer reflect the current beer in the barrel.[12] Many of the brewers and blenders I spoke with shared variations on this theme. Most do not look back at previous notes when returning to a given beer and also do not track the progression of the beer over time. Some take this to the extreme, not recording any notes at all when tasting!

If just getting started, you should probably err on the side of taking too many notes rather than too few. Remember, many of the blenders I interviewed have been successfully managing their programs for years and have a deep understanding of how each of their beers develops over time. Initially, while learning how your beers change as they mature you may find significant value in tracking the way the flavors of each barrel evolve over time, potentially allowing you to identify

10 A. Zinn, telephone conversation with author, May 22, 2020.
11 Swanson, interview.
12 H. McCabe, telephone conversation with author, May 4, 2020.

Figure 14.2. Foeders are vertical or horizontal tanks made from wood. © Getty/Mmaxmax

trends or early indications of specific flavors that may arise in the mature beer. However, as you learn and internalize this information, do not be afraid to stop collecting data that no longer serves you. Ultimately, you should aim to streamline your note-taking and data collection to include only the elements that help you continue making great beer.

Providing Tasters with Information

In the formal sensory setting, we usually present beers to panelists with as little identifying information as possible to avoid biasing their evaluations. In TTT testing we only provide panelists with the brand identity; in most description tests we give them even less, often providing no identifying information at all. When tasting barrel-aged beer it is more challenging to limit information. Most tastings will only involve one or two tasters, and the barrels tasted have to be identified in advance and sampled from directly, often by those same tasters. However, if one person prepares all of the samples, you can decide exactly how much or how little information you want the other tasters to have. Some breweries will present all relevant batch

data, including the recipe, age, and type of barrel. Other breweries will opt to present information in a manner more akin to a formal sensory tasting, providing the identity of the beer with no other details.

In the end, it is up to you to decide how much you want to tell tasters about the barrel-aged samples that they taste. Just remember that, as with any other tasting, giving tasters information about the beer provides them with knowledge that may bias their evaluation. Given the variable nature of wood-aged beers and the rate at which they change over time, I recommend against providing tasters with access to tasting notes from previous sessions prior to assessing a beer. If you allow tasters to see their former tasting notes before they evaluate the beer, their past impressions will heavily influence their present assessment.

BLENDING BEER

While most blenders fell into one of a few different camps when it came to methods for tracking wood-aged beer, no two blenders were exactly alike when it came to blending the beer itself. A number of different factors influence blending decisions and methods, including the size of the program, the type of beer being made, the goals of the brewery, and the blender's philosophy. The following section covers a variety of different perspectives and philosophies from a wide selection of breweries and blenders. Depending on the specific goals of your own program, pick the pieces that you think might work best for you and try incorporating them into the way you blend wood-aged beer.

Different Approaches to Blending

Vintage Approach

Most blenders of wood-aged beer approach the overall philosophy of their program with a looser definition of a brand than we see in the traditional, consistency-focused sensory environment. Now, this is not true in all cases. Especially with larger producers that send their beers into distribution, consistency in their wood-aged beers often matters every bit as much as it does in any other beer they make, with these beers appearing on their standard product release TTT panel following blending. However, most small- to medium-sized programs take more of a "vintage" approach to their brands. This means accepting—and in some cases even expecting—some amount of variation from blend to blend. When preparing to assemble a new blend of an existing brand, Cory King of Side Project Brewing shared that he typically begins the process by pulling bottles of previous blends to taste critically. In tasting previous iterations of a given brand, he identifies what he likes best about that beer and what elements he wants to recreate, but also considers ways he can potentially improve on it with the upcoming blend.[13]

Blending via Elimination

In larger programs, blenders can produce highly consistent brands by blending via elimination. While he approaches new projects differently, Andy Parker will typically blend Avery's well-established barrel-aged brands via elimination, tasting each barrel of a given batch and eliminating barrels that do not fit the blend. Barrels might get kicked out for exhibiting certain off-flavors, or simply because they do not match the overall profile that he is looking for in the finished blend.[14] When blending beer using this method, test blends do not offer much value and are not often used.

Build a Base, then Tweak

Many blenders making medium to large batches (probably a minimum of 8–10 barrels blended together) approach blending by building a base of similar beer and then seasoning the blend with a few characterful barrels. At The Bruery, Jeremy Grinkey usually likes to begin by building a cohesive base before selecting a couple of fun, interesting pieces to drive complexity in the finished beer. As an example, he recounted blending a blond peach sour and adding a couple of barrels to the blend that presented strong floral aromas of geranium. While the assertive floral character might have overtaken the stone fruit if it were the sole base used for the blend, the addition of a couple of barrels to a larger base blend lent a unique, subtle herbal note to the finished beer. In describing this beer, Grinkey noted that complexity in wood-aged beer is not driven by the things that are perfect, but rather by the things that are imperfect.[15]

Andrew Zinn shared a similar process, using tasting notes to divide his cellar into barrels that he can use to build a base and more distinctive "character" barrels. Zinn explains that, in blending, it is not about simply combining all of your most exceptional barrels into a finished beer—a blend made entirely of character barrels will likely come out as a muddy, directionless mess. Instead, he typically selects a few character barrels that he feels will mesh well together, and then layers those over a more neutral base to allow those character barrels to shine.[16]

Brewers who approach blending using this method nearly universally use test blends to refine the exact blend prior to racking the barrels together. Methods for assembling test blends differ. Some brewers will mix using graduated cylinders, or even a syringe or micropipette for high levels of accuracy. Some brewers simply free pour relatively equal amounts of the different samples into a single cup.

Many brewers will assemble several test blends and pit them against one another to see which they prefer, with some going as far as running basic hedonic testing on

---

[13]   C. King, telephone conversation with author, April 20, 2020.
[14]   Parker, telephone conversation.
[15]   Grinkey, telephone conversation.
[16]   Zinn, telephone conversation.

the different test blends with a wider group of tasters to see which one scores the highest. Some brewers will also use test blends multiple times throughout the process of blending, perhaps building a base of 80% of the total blend volume and tasting that base to determine what characteristics they need to tweak with the final 20%. Regardless of which method appeals to you, test blends offer an indispensable tool for evaluating different combinations of barrels before committing to a final blend.

CAN YOU BLEND WITH PARTIAL BARRELS?

For a new program or small-scale blends, you might wonder whether it is possible to only use part of a barrel in a blend. Of the blenders I spoke to who age their beer in barrels or puncheons (vessels with a capacity between 200 and 600 liters), I did not encounter anyone who uses partial barrels to construct a blend. Leaving a barrel half full greatly increases the head space in the vessel, which allows a significant amount of oxygen to come into contact with the liquid, causing any remaining beer to rapidly develop high levels of vinegary acetic acid. For programs using larger-format foeders, responses were mixed: some brewers would draw beer off of a foeder for blending before topping the foeder up with fresh beer, while others would empty entire foeders into a single blend, much like most breweries do with individual barrels. If you are trying to blend with a potently flavored barrel and think that adding the whole barrel will overtake the blend, you can resort to using only a portion of that barrel—just know that you will likely have to use the remaining liquid in short order or dump it.

Small-Scale Blending

Small-scale blending involves blending a handful of barrels together to produce a batch. At this scale, some brewers still use the approach of building a base around a single character barrel, but some brewers will instead

look for each barrel to contribute unique aspects to the blend. When she led the program at Jester King, Averie Swanson would often look for each barrel of a blend to show one prominent feature, such as a higher level of acidity or bitterness, or pronounced fruit or *Brettanomyces* fermentation character. While producing a large batch by randomly blending barrels with different traits together would likely lead to a messy, unfocused finished beer, the smaller batch size allowed Averie to carefully select barrels so that the prominent features of each individual barrel were balanced against one another in the finished blend.[17] At this scale, using test blends is essential to make sure that the different elements do, in fact, find balance in the finished blend.

Blake Tyers of Creature Comforts Brewing Company often approaches blending through the lens of trying to build a single flavor idea from several different angles. Much like an orchestra builds a single rich, layered chord sound by using instruments with differing timbre or tone color, Tyers can build a complex impression of a single flavor by layering different potential sources. As an example, he discussed building orange flavor from orange zest, hops, flowers, and fermentation, but the concept can be applied across the flavor spectrum. When building a blend, he begins with an idea in mind of what he wants to create, and then selects barrels from his stock that express those desired characteristics in slightly different ways to produce a richly textured finished product.[18]

Blending against Added Ingredients

When blending wood-aged beers that have fruit or adjuncts added, brewers will often attempt to balance the blend of the underlying base beer against the added ingredients. With fruit going into a wild beer, the balancing act often begins by considering the desired acid level in the finished beer. Jason Perkins explains that with fruits like sour cherries and apricots, which bring a substantial amount of their own acid to the beer, he often tries to build a base beer with lower acidity so that he does not end up with a bracingly sour finished beer.[19]

In addition to acidity, brewers will often also consider how the flavor and structural elements of the base blend will work with the added fruit. With a large enough stock, blenders can build a base beer that will

17   Swanson, interview.
18   B. Tyers, telephone conversation with author, April 21, 2020.
19   Perkins, telephone conversation.

support or complement the flavors of the fruit. For example, Jeremy Grinkey noted that The Bruery's sour blonde base stock can develop a handful of different flavors as it ages—from stone fruit to horse blanket to kalamata olives—all of which can be positive characteristics in the right context. However, if he is trying to make a peach beer, it is a lot easier to build off of a base that already shows some stone fruit character rather than a bunch of barrels that taste of olives.[20]

Brewers will also take into account the characteristics that the fruit itself expresses. When consumers think about peaches or oranges, they typically have a single characteristic expression of the fruit in mind. However, fruits can express a wide variety of differing sugar levels, acid levels, and primary flavor notes depending on the specific variety of the fruit (think of cabernet sauvignon vs. chardonnay grapes, or a Honeycrisp apple vs. a Granny Smith) as well as a variety of terroir-type factors such as soil, location, climate, and weather, the latter of which can further vary from year to year. Prior to using fruit in beer, you should evaluate the fruit, much like you would any of your raw materials. In fact, some brewers will perform this sort of evaluation before even purchasing their fruit. Jim Crooks of Firestone Walker Barrelworks treats fruit selection in the same way that many brewers treat hop selection, tasting and evaluating the characteristics of a batch of fruit and thinking about whether it will work in his beer before making a purchase.[21]

A final consideration, important with both fruit and other adjuncts, is how much to add to the finished beer. Some brewers will simply use a predetermined amount, for example, always adding strawberries at three pounds per gallon for a certain brand. Some brewers will add an excess of fruit to the base and then will blend in beer with no added fruit to taste, adding it until they achieve the level of fruit character they are looking for. Others still will use test blends to experiment with different levels of added fruit to determine the perfect level. Ron Jefferies explains that he typically tests three to five different levels of added fruit for each fruit that he uses in a beer. This can quickly become tedious. Ron notes that if adding two fruits to a beer it can lead to anywhere from 10 to 25 different test blends to evaluate

each combination of the different fruit levels. However, he finds that this allows him to target the exact profile he is looking for, adjusted to account for the specific batch of fruit he is working with, which makes all of the trial and error worthwhile.[22]

Marty Scott at Revolution describes a similar process when adding adjuncts to stouts, such as coffee or vanilla, among others. However, he notes that sometimes the brewers use a packaged version of the same beer as the base for their test batches rather than using the actual batch of beer that will have adjuncts added. This might not work as well with wild beer but, given that clean, spirit barrel–aged beer tends to develop more predictably, this can be done with limited risk in certain situations. In Scott's experience, it is more important to taste beer that is cold and carbonated than it is to taste the exact wood-aged base beer blend when attempting to determine desired adjunct levels.[23]

Assemblage versus Coupage

While talking to Frank Boon about how he blends lambic, he introduced me to two different schools of thought when it comes to blending beer: *assemblage* and *coupage*. Assemblage describes the assembly of different barrels of beer—all of which are acceptable in their own right—into a finished product that is better than the sum of its parts. On the other hand, coupage, which is French for "cutting," describes the process of taking a substandard barrel of beer and trying to blend the rough edges or negative characteristics away, so as not to waste that beer. At Brouwerij Boon, assemblage is the standard. Given Frank's extensive experience in making lambic beer, he rarely has a foeder turn on him. On the rare occasion that this happens, the brewery sells the liquid to a distillery rather than attempt to hide it in a large batch of lambic.[24]

However, a single barrel of beer that might be considered bad on its own can sometimes be a useful tool in the blender's arsenal. I spoke with several brewers who describe how they use beer that they would never sell on its own to add special or unique characteristics to a finished blend. Jason Perkins gave the example of a four-year-old barrel with loads of

---

20   Grinkey, telephone conversation.
21   J. Crooks, telephone conversation with author, May 7, 2020.
22   Jeffries, telephone conversation.
23   Scott, telephone conversation.
24   Boon, telephone conversation.

sherry oxidation that yielded nuanced aged character in a larger blend.[25] Ron Jefferies went a step further, describing how occasionally you might find a barrel of beer so awful that you want to burn the actual barrel itself, but which could add something really special to a 100-barrel blend.[26]

While these anecdotes colorfully illustrate that a "bad" barrel can be used to good effect, they do come with a few caveats. First, these brewers used highly characterful barrels to season a large blend of beer, not as a significant portion of the final blend. This does not really qualify as coupage. In both Jason's and Ron's cases, their goal was not to try to build a blend to salvage some bad beer, but rather to use an intensely flavored beer to add a dimension of complexity to a blend. Second, cases such as these absolutely require assembly of test blends. These brewers are not simply adding such an intensely flavored barrel to a large blend and hoping that it works out. While their years of experience led them to believe that such an addition *might* make for an interesting finished beer, they still tested this theory prior to blending to avoid potentially ruining the entire batch.

Dumping Beer

The brewers that I talked to universally acknowledged that production of wood-aged beer results in a significantly higher dumping rate compared to standard beer production. If you are just beginning a barrel program, go in with the expectation that you will have to dump some beer. As previously discussed, certain undesirable or overly assertive flavors can add pleasant complexity to a blend, but some flavors are not worth blending into a finished beer.

Nearly all producers of wild wood-aged beers mentioned ethyl acetate and acetic acid as the most common issues that led to a barrel being dumped. While low levels of either of these characteristics can sometimes accentuate fruit character in a sour fruit beer, higher levels quickly become solventlike (ethyl acetate) or vinegary (acetic acid). Other assertive off-flavors—such as trichloroanisole (TCA) in any beer, or *Enterobacter* flavors in lambic—indicate mold or spoilage and typically lead to prompt removal of the beer and the vessel itself to avoid contaminating other barrels in the cellar. With clean wood-aged beer, any notable acidity, which indicates microbial contamination, similarly leads to removal of both beer and barrel.

Tasting Warm, Flat Beer

One of the difficulties inherent in blending wood-aged beer is working with samples that differ dramatically from the way the finished beer will present. You will use your sensory acuity to build and evaluate test blends assembled from warm, uncarbonated barrel stock. However, your goal is not necessarily to produce a delicious test blend, but rather a test blend that will transform into something delicious once it is cold, carbonated, and packaged. Learning to work backward from your desired flavor profile to a corresponding test blend is a skill mastered only through time and experience. Cory King quipped that he tastes more flat beer than finished beer, a testament to the amount of tasting required to produce these blends.[27] In discussing her thought process, Averie Swanson says she usually tries to envision how carbonation will bring certain flavors to the forefront while muting others, and how the level of carbonation will impact mouthfeel and bitterness.[28]

For new or special projects, Andy Parker does not force his tasters to extrapolate how the finished blend will taste when cold and carbonated. Instead, he goes to the trouble of cooling his test blends in two-liter soda bottles and then force carbonating them using a carbonator cap. While Andy admits that the process is tedious and time-consuming, this extra step makes sensory evaluation of test blends a lot more straightforward. It also allows him to serve samples to a wider group of tasters that do not necessarily have sufficient experience to imagine how an uncarbonated, room temperature blend will taste when finished.[29]

For brewers who bottle-condition their wood-aged beer, the carbonation step offers a final opportunity to tweak the flavor profile. Unsurprisingly, decisions at this point are usually made based on sensory cues. At Fonta Flora Brewery, Jeremy Inzer explains how they have different yeasts that they use

25 Perkins, telephone conversation.
26 Jeffries, telephone conversation.
27 King, telephone conversation.
28 Swanson, interview.
29 Parker, telephone conversation.

for bottle conditioning depending on the profile of the beer. If the brewers are completely satisfied with the state of the beer and do not want any additional flavor added from the bottle refermentation, they opt for a neutral champagne yeast. However, if they want to increase either fruity notes or *Brettanomyces* character, they have their own house mixed-culture blends that they add prior to packaging to make minute, final adjustments.[30]

Following bottle conditioning, brewers typically monitor the beers over the course of weeks or even months, waiting for carbonation to develop and certain undesirable flavors to subside. In the case of wild beers, tetrahydropyridine (THP) can be a particularly pernicious flavor. For many producers of wild beers, the eventual reduction of THP serves as the signal to release the beer. At Oxbow Brewing Company, Mike Fava says the mixed fermentation beers spend a minimum of two months conditioning before they even come to the table for a final quality control check. If the beer still shows signs of THP, Oxbow's tasters will retest the beer every two weeks, only releasing it when those flavors have subsided.[31]

FINDING YOUR PHILOSOPHY

Over time, you will begin to build your own personal blending philosophy based on your approach, your techniques, and your goals. And while you can learn a lot from the methods of world-class brewers such as those interviewed for this chapter, there is no substitute for experience, particularly when it comes to understanding the sensory elements that make for a good finished blend. Jay Goodwin of The Rare Barrel recommends leaning on the collaborative nature of our industry and trying to find an experienced blender to work with you for your first few batches. While this may require you to swallow your pride and admit that you need help, you cannot simply read your way to becoming a good blender.[32] Asking someone with experience to assist you can help you avoid potential pitfalls with your first few blends.

To develop your skills, Blake Tyers suggests taking meticulous notes on your blends. If you have a small

barrel program, you will not get that many shots at blending throughout the course of a year, meaning fewer opportunities to learn from your mistakes and build on your successes. Additionally, if you bottle-condition your beer there can be a significant delay in receiving feedback—you may not know how your blend has actually turned out until three or more months after you put it together. Taking thoughtful and thorough sensory notes on your test blends and then comparing these notes against the finished beer makes it easier for you to refine your process, even if you do not have frequent opportunities to practice.[33]

As a final word of wisdom, do not limit yourself when it comes to blending wood-aged beer. Cole Hackbarth of Rhinegeist Brewery noted that they often mix non-wood-aged beer into their blends prior to packaging to achieve the characteristics that they desire in the finished beer. As beer ages in wood it can extract tannins, leading to a thinner and more astringent mouthfeel. In a beer like Rhinegeist's Ink Imperial Stout, they are not always able to achieve the desired rich, chewy body in the finished beer using only beer aged in wood. To solve this problem, Rhinegeist often blends either fresh Ink or aged (but not wood-aged) Ink with the wood-aged beer to yield a finished beer that has the flavor characteristics of wood-aging alongside the intended mouthfeel.[34]

Jim Crooks goes even further in this regard, as illustrated by his description of the blending process for a 200-barrel batch of Bretta Rosé, Firestone Walker's take on a raspberry Berliner weisse. Usually, Jim will put together 190 barrels of the batch before using the last 10 barrels as "salt and pepper" to adjust and perfect the blend. At that point, virtually anything is on the table. Yes, he could pull other barrel-aged beer from a variety of different types of stock, but if the blend tastes too sour, he might reach for 10 barrels of the brewery's 805 Blonde Ale or Firestone Lager. Jim notes that, while he has never actually had to resort to this, even using water to simply dilute an overly intense batch is something he has considered.[35]

In speaking with Jim Crooks, he closed our conversation by explaining that similar logic or

30   J. Inzer, telephone conversation with author, May 7, 2020.

31   M. Fava, telephone conversation with author, May 5, 2020.

32   J. Goodwin, telephone conversation with author, April 21, 2020.

33   Tyers, telephone conversation.

34   C. Hackbarth, telephone conversation with author, May 12, 2020.

35   Crooks, telephone conversation.

techniques could be applied to a much smaller barrel program as well. Perhaps your brewery only has five barrels of wood-aged beer to work with, but how many other brands and types of beer do you have available to you? In blending, the ultimate goal is to produce a finished product better than the sum of its original parts. Do not allow a small stock of wood-aged beer to constrain you. Leverage the tools at your disposal and allow your senses to guide you toward blending better beers.

15
THE ROAD AHEAD

If you made it this far, you are now equipped with everything you need to know to get your sensory program off the ground. You have built a foundation of knowledge that covers sensory systems, sampling techniques, and beer flavor. You know the methods most commonly used in sensory work—you know how to apply those methods and which methods are best suited for which scenarios. Most importantly, you know how to get started. Now begins the work of actually building your program. If you are thoughtful and intentional about the way you train your panelists and run your panel sessions, you will find yourself with a tool that will serve your brewery for years to come.

Keep the key principles highlighted in the first chapter in mind whenever designing tests or running panels (see pp. 10–11). Frame each test as a means to address a specific question. Design each test to yield data that allows you to act or make decisions. Build action standards before performing your tests so that you know how to proceed after analyzing your results. Use quick tests like the true-to-target test to help you identify when further testing may be required. And be sure to communicate with other departments to make sure that the data generated by your panelists actually gets used to help the brewery make better beer. With time and experience, you will develop an intuitive understanding of how the key principles can be applied to solve new and different questions beyond those laid out in these pages. And as your program begins to shine within the brewery, you will likely be asked broader questions by other departments, questions that may require the creative application of these principles.

If you want to develop your program beyond the content of this book, one opportunity for growth is to increase the scope of your training program. The options for additional training are nearly endless, but, as always, make sure that any further investment of resources offers the brewery a tangible benefit. You could simply train your panelists on a wider pool of specific attributes, though other skill sets may offer equal, or even greater, value. Expand your flavor training to help your panelists grow their lexicon and become familiar with a wider array of possible beer descriptors. You can also increase the amount of descriptive profiling your panel does to include products other than those produced by your brewery. Descriptive profiling can be used to better understand beers made by other breweries, as well as products beyond beer, such as coffee, chocolate, or whiskey. If your brewery wants to produce a beer that mimics the flavors of whiskey, you can use your panelists to profile a group of different whiskeys to identify the flavors that you want to target in the beer. As your panelists gain experience and broaden their skills, your sensory program will naturally come to play a larger role in product development.

It is certainly exciting to consider what possibilities the future may hold, but first you need to build your program. You may have to make some concessions at the beginning—perhaps you may not have as many panelists as you would like, or you might not have the perfect space—but do not let that keep you from getting started. Great sensory programs are not built overnight; your program does not have to be (and honestly, will not be) perfect from day one. At this stage, keep in mind your central objective—releasing quality beer that tastes the way you want it to. This goal is accessible and entirely within your grasp. Now get out there and get started!

APPENDIX A
RANK-SUM TEST STATISTICAL TABLES

Rank-sum test statistical tables on following page.

After computing the rank sums for a ranking test, calculate the differences between each pair of rank sums. Match the total number of assessors (N) with the column indicating the number of samples presented in the test. Compare this value to the differences between each pair of rank sums. If any of the differences exceed the value in the table, the relative ordering of those two samples can be considered statistically significant.

Table A.1 **Ranking Test – Critical values for the differences between rank sums***

	Number of Samples						
N	3	4	5	6	7	8	9
3	6	8	11	13	15	18	20
4	7	10	13	15	18	21	24
5	8	11	14	17	21	24	27
6	9	12	15	19	22	26	30
7	10	13	17	20	24	28	32
8	10	14	18	22	26	30	34
9	10	15	19	23	27	32	36
10	11	15	20	24	29	34	38
11	11	16	21	26	30	35	40
12	12	17	22	27	32	37	42
13	12	18	23	28	33	39	44
14	13	18	24	29	34	40	46
15	13	19	24	30	36	42	47
16	14	19	25	31	37	42	49
17	14	20	26	32	38	44	50
18	15	20	26	32	39	45	51
19	15	21	27	33	40	46	53
20	15	21	28	34	41	47	54
21	16	22	28	35	42	49	56
22	16	22	29	36	43	50	57
23	16	23	30	37	44	51	58
24	17	23	30	37	45	52	59
25	17	24	31	38	46	53	61
26	17	24	32	39	46	54	62
27	18	25	32	40	47	55	63
28	18	25	33	40	48	56	64
29	18	26	33	41	49	57	65
30	19	26	34	42	50	58	66

*The values in this table are derived from Kramer's rank-sum test and correspond to $\alpha = 0.05$.
Reference: Harry T. Lawless and Hildegarde Heymann, *Sensory Evaluation of Food: Principles and Practices*, 2nd ed. (New York: Springer Science+Business Media, 2010), 563.

APPENDIX B
TRIANGLE TEST AND TETRAD TEST STATISTICAL TABLES

Triangle test and tetrad test statistical tables on following page.

After selecting values for α, β, and P_d based on the particulars of the test you will be running, use this table to determine the minimum number of assessors you will need to perform a triangle test.

Table B.1 **Triangle Test – Minimum Number of Assessors Needed**

P_d	α	β 0.20	0.10	0.05	0.01	0.001
50%	0.20	7	12	16	25	36
	0.10	12	15	20	30	43
	0.05	16	20	23	35	48
	0.01	25	30	35	47	62
	0.001	36	43	48	62	81
40%	0.20	12	17	25	36	55
	0.10	17	25	30	46	67
	0.05	23	30	40	57	79
	0.01	35	47	56	76	102
	0.001	55	68	76	102	130
30%	0.20	20	28	39	64	97
	0.10	30	43	54	81	119
	0.05	40	53	66	98	136
	0.01	62	82	97	131	181
	0.001	93	120	138	181	233
20%	0.20	39	64	86	140	212
	0.10	62	89	119	178	260
	0.05	87	117	147	213	305
	0.01	136	176	211	292	397
	0.001	207	257	302	396	513
10%	0.20	149	238	325	529	819
	0.10	240	348	457	683	1011
	0.05	325	447	572	828	1181
	0.01	525	680	824	1132	1539
	0.001	803	996	1165	1530	1992

Reference: Lauren Rogers, ed., *Discrimination Testing in Sensory Science: A Practical Handbook* (Duxford, UK: Woodhead Publishing, 2017), 474.

Match the total number of assessors used (N) with the α value selected for your test to find the minimum number of responses required to declare that a statistically significant difference exists. Declare that a difference exists if the number of correct responses is equal to or exceeds the number shown in the table.

Table B.2 **Tetrad Test - Minimum Number of Assessors Needed**

δ	α	β				
		0.20	0.10	0.05	0.01	0.001
0.50	0.20	367	560	759	1204	1842
	0.10	560	806	1028	1557	2274
	0.05	752	1029	1293	1873	2649
	0.01	1202	1544	1867	2551	3451
	0.001	1822	2251	2636	3444	4476
0.75	0.20	88	133	176	277	414
	0.10	132	185	236	350	508
	0.05	173	234	293	420	594
	0.01	272	349	418	569	769
	0.001	411	501	586	766	995
1.00	0.20	35	49	68	102	156
	0.10	50	69	88	132	188
	0.05	65	89	110	154	220
	0.01	101	130	156	210	283
	0.001	152	183	214	280	365
1.25	0.20	19	27	35	49	74
	0.10	24	32	45	66	91
	0.05	34	42	52	78	105
	0.01	51	61	76	101	135
	0.001	75	92	102	135	175
1.50	0.20	11	16	19	30	43
	0.10	15	20	25	37	53
	0.05	20	25	32	42	60
	0.01	30	37	46	57	76
	0.001	43	54	61	78	100

Reference: Lauren Rogers, ed., Discrimination Testing in Sensory Science: A Practical Handbook (Duxford, UK: Woodhead Publishing, 2017), 479-80.

After selecting values for α, β, and δ based on the particulars of the test you will be running, use this table to determine the minimum number of assessors you will need to perform a tetrad test.

Table B.3 **Triangle Test or Tetrad Test – Minimum Number of Correct Responses Needed to Conclude a Statistically Significant Difference Exists**

N	α				
	0.2	0.1	0.05	0.01	0.001
15	8	8	9	10	12
16	8	9	9	11	12
17	8	9	10	11	13
18	9	10	10	12	13
19	9	10	11	12	14
20	9	10	11	13	14
21	10	11	12	13	15
22	10	11	12	14	15
23	11	12	12	14	16
24	11	12	13	15	16
25	11	12	13	15	17
26	12	13	14	15	17
27	12	13	14	16	18
28	12	14	15	16	18
29	13	14	15	17	19
30	13	14	15	17	19
32	14	15	16	18	20
34	15	16	17	19	21
36	15	17	18	20	22
38	16	17	18	20	23
40	17	18	19	21	24
42	18	19	20	22	25
44	18	20	21	23	25
46	19	20	22	24	26
48	20	21	22	25	27
50	20	22	23	25	28
55	22	24	25	27	30
60	24	26	27	30	33
65	26	28	29	32	34
70	28	29	31	34	37
75	29	31	33	35	39
80	31	33	35	37	41
85	33	35	36	39	43
90	35	37	38	42	45
95	37	39	40	43	47
100	38	40	42	45	49

To calculate the number of correct responses necessary (X) for a number of panelists (N) not listed here, use the following formula*: $X = \left(\frac{2N+3}{6}\right) + Z\sqrt{\frac{2N}{9}}$

*Z corresponds to the value selected for **α**: =0.20, Z=0.84; =0.10, Z=1.28; =0.05, Z=1.64; =0.01, Z=2.33; =0.001, Z=3.09.

Reference: M.C. Meilgaard, Gail Vance Civille, and B. Thomas Carr, Sensory Evaluation Techniques, 5th ed. (Boca Raton, FL: CRC Press, 2016), 554.

APPENDIX C
SAMPLE BALLOTS FOR SENSORY TESTS

This appendix contains sample ballots for the different sensory tests covered throughout the book. If you decide to create your own sensory ballots rather than using a sensory software program, you can use these examples to build your ballots in Google Forms or some other medium of your choice.

PRODUCT RELEASE TRUE-TO-TARGET (TTT) TEST

The ballot shown below uses the witbier profile presented in figure 10.1 (p. 120). To create your own ballots, replace the descriptions with the corresponding descriptions for your own brands. When this test is presented to panelists, you should indicate on the ballot that this is a product release TTT test.

Product Release True-To-Target (TTT) Test Ballot

Panelist Instructions: This is a product release sample of *Belgian Witbier.* Assess each modality against the target description. For each modality, please indicate whether the modality is true to target (TTT) or not true to target (Not TTT). For any modalities that you mark as Not TTT, please describe how the sample differs from the target description. You may also provide feedback on modalities marked TTT if desired. After assessing each modality, make a final assessment of whether the sample overall is TTT or not TTT.

Target Appearance: Pale straw color, medium haze, white foam, high foam retention

    ☐ TTT   ☐ Not TTT

Feedback:

Target Aroma: Coriander, lemongrass, and orange peel, with slight black peppercorn, white bread, and grainy aromas

    ☐ TTT   ☐ Not TTT

Feedback:

Target Taste: Low bitterness, low sweetness, no sourness

    ☐ TTT   ☐ Not TTT

Feedback:

Target Mouthfeel: Medium body, high carbonation, no astringency, no alcohol warmth

    ☐ TTT   ☐ Not TTT

Feedback:

Overall Target: Must feature pale straw color, medium haze, coriander aroma, low bitterness, and high carbonation. Cannot contain perceivable sourness or vegetal celery notes

    ☐ TTT   ☐ Not TTT

Feedback:

SHELF LIFE TRUE-TO-TARGET (TTT) TEST

This ballot uses the witbier profile presented in chapter 10 (p. 120). To create your own ballots, replace the descriptions with the corresponding descriptions for your own brands. When this test is presented to panelists, you should indicate on the ballot that this is a shelf life TTT test.

Shelf Life True-to-Target (TTT) Test Ballot

Panelist Instructions: This is a shelf life sample of *Belgian Witbier*. Assess each modality against the target description for a fresh sample of *Belgian Witbier*. For each modality, please provide feedback on ways in which the beer has changed, specifically commenting on whether the original traits have increased, decreased, or remained the same and noting any new flavors or characteristics that have emerged. After entering your feedback, indicate whether that modality is true to target (TTT) or not true to target (Not TTT). After assessing each modality, make a final assessment of whether the sample overall is TTT or Not TTT.

Target Appearance: Pale straw color, medium haze, white foam, high foam retention

    ☐ TTT   ☐ Not TTT

Feedback:

Target Aroma: Coriander, lemongrass, and orange peel, with slight black peppercorn, white bread, and grainy aromas

    ☐ TTT   ☐ Not TTT

Feedback:

Target Taste: Low bitterness, low sweetness, no sourness

    ☐ TTT   ☐ Not TTT

Feedback:

Target Mouthfeel: Medium body, high carbonation, no astringency, no alcohol warmth

    ☐ TTT   ☐ Not TTT

Feedback:

Overall Target: Must feature pale straw color, medium haze, coriander aroma, low bitterness, and high carbonation. Cannot contain perceivable sourness or vegetal celery notes

    ☐ TTT   ☐ Not TTT

Feedback:

BRIGHT TANK SPOT CHECK TRUE-TO-TARGET (TTT) TEST

While this test can be performed without a ballot, it is useful to require tasters to quickly fill out a ballot to focus their attention on this step. Since this ballot only features an assessment of the overall target, make sure your overall target description captures the appearance of the beer—appearance typically offers the simplest check to make sure that the correct beer is being packaged. This ballot uses the witbier profile presented in chapter 10 (p. 120). To create your own ballots, replace the descriptions with the corresponding descriptions for your own brands.

Bright Tank Spot Check True-to-Target (TTT) Test Ballot

Panelist Instructions: This is a bright tank spot check of *Belgian Witbier.* Assess the beer against the overall target description and indicate whether the beer is true to target (TTT) or not true to target (Not TTT). If you mark the beer as Not TTT, provide feedback **and do not package the beer!!**

Overall Target: Must feature pale straw color, medium haze, coriander aroma, low bitterness, and high carbonation. Cannot contain perceivable sourness or vegetal celery notes

☐ TTT   ☐ Not TTT

Feedback:

ACCEPTANCE (HEDONIC) TEST

If using the acceptance test to evaluate a single sample, you can label the sample however you wish, or even present the sample without a label at all. If you are presenting multiple samples for acceptance testing within a single session, you may want to consider using random, 3-digit numbers to label your samples to avoid biasing your panelists.

Acceptance (Hedonic) Test Ballot

Panelist Instructions: Please taste Sample #1. After evaluating the sample, select one of the following phrases to indicate your overall opinion of the sample.

☐ Like Extremely

☐ Like Very Much

☐ Like Moderately

☐ Like Slightly

☐ Neither Like Nor Dislike

☐ Dislike Slightly

☐ Dislike Moderately

☐ Dislike Very Much

☐ Dislike Extremely

PREFERENCE TEST

With preference tests, most texts recommend using random, 3-digit numbers to label your samples to avoid biasing panelists.

Preference Test Ballot

Panelist Instructions: Please taste the two samples in the order presented. After evaluating both samples, you may revisit either of the samples if you wish. Once you have finished evaluating the samples, please mark the box next to the sample number of the sample that you prefer. You must select one of the samples—an answer of "no preference" is not permitted.

☐ Sample 745   ☐ Sample 884

RANKING TEST

Ranking tests can be used as an affective test to rank samples based on their hedonics (i.e., how much your panelists like them) or as a descriptive test to rank samples based on the intensity of a given trait. With this type of test, most texts recommend using random, 3-digit numbers to label your samples to avoid biasing panelists. Examples of both types of test are presented below. In the case of the descriptive test, "citrus aroma" can be replaced with any attribute by which you would like to sort the samples.

Affective Test

Affective Test Ballot

Panelist Instructions: Please taste the four samples in the order presented. After evaluating all of the samples, you may revisit any of the samples if you wish. Rank the samples from most preferred to least preferred using numbers 1 to 4 in the following order: 1–most preferred; 4–least preferred. Each sample must be given a distinct rank—no ties are allowed.

Sample Rank (1–4)

684 _____

923 _____

415 _____

162 _____

Descriptive Test

Descriptive Test Ballot

Panelist Instructions: Please taste the five samples in the order presented. After evaluating all of the samples, you may revisit any of the samples if you wish. Rank the samples from lowest to highest intensity of *citrus aroma* using numbers 1 to 5 in the following order: 1–least intense; 5–most intense. Each sample must be given a distinct rank—no ties are allowed.

Sample Rank (1–5)

224 _____

671 _____

958 _____

853 _____

118 _____

DESCRIPTION TEST

The description test can be used to build a descriptive profile for a beer or a raw material. The following ballot covers some of the most common attributes that you might track for a beer, but the exact attributes you choose to cover are up to you. For a full list of potential attributes, see chapter 4. For aroma assessment, if your panel uses a lexical set other than the Beer Flavor Map, you can replace mention of the Beer Flavor Map within the aroma section on your own ballot.

Description Test Ballot

Panelist Instructions: Evaluate Sample #1 and record your assessment of each attribute below.

Appearance:

Color: Record the color using one of the following descriptors: straw, yellow, orange, amber, brown, black. You can apply the words "light" or "dark" to any of those descriptors if desired.

Haze/Turbidity:	None	Low	Medium	High

Aroma:

Record the aromas you perceive using terms from the *Beer Flavor Map*:

_____	_____	_____
_____	_____	_____
_____	_____	_____

Taste:

Sweetness:	None	Low	Medium	High
Bitterness:	None	Low	Medium	High
Sourness:	None	Low	Medium	High

Mouthfeel:

Body:		Low	Medium	High
Carbonation:		Low	Medium	High
Alcohol Warmth:	None	Low	Medium	High
Astringency:	None	Low	Medium	High

SINGLE ATTRIBUTE SCALING TEST

In most cases, this test would be used to compare several samples based on the intensity of a single attribute. For that type of test, most texts recommend using random, 3-digit numbers to label your samples to avoid biasing panelists. Note that "sour taste" can be replaced with any attribute that you would like your panelists to evaluate, and you can use a scale with a different number of points (e.g., 7-point, 9-point, etc.) if you would like.

Single Attribute Scaling Test Ballot

Panelist Instructions: Taste sample 553 and rate the intensity of *sour taste* using the following scale: 1–low intensity; 5–high intensity. Rate the sample using an integer value (i.e., 1, 2, 3, 4, or 5); decimal ratings are not allowed.

Sour taste intensity rating for sample 553: _____

TRIANGLE TEST

With difference tests, most texts recommend using random, 3-digit numbers to label your samples to avoid biasing panelists.

Triangle Test Ballot

Panelist Instructions: Of the three samples in front of you, two samples are the same, and one is different. Please taste the samples in the order presented to you, from left to right. After evaluating all three samples, please circle the number of the sample that is different.

| Sample #490 | Sample #996 | Sample #143 |

TETRAD TEST

With difference tests, most texts recommend using random, 3-digit numbers to label your samples to avoid biasing panelists.

Tetrad Test Ballot

Panelist Instructions: Of the four samples in front of you, two samples belong to one group, and two samples belong to a different group. Please taste the samples in the order presented to you, from left to right. After evaluating all four samples, please group the samples into two groups of two based on similarity.

Group 1 (Please write in sample codes)   Group 2 (Please write in sample codes)

_____ _____   _____ _____

BIBLIOGRAPHY

Adams, An and Norbert De Kimpe. 2006. "Chemistry of 2-Acetyl-1-pyrroline, 6-Acetyl-1,2,3,4-tetrahydropyridine, 2-Acetyl-2-thiazoline, and 5-Acetyl-2,3-dihydro-4H-thiazine: Extraordinary Maillard Flavor Compounds." *Chemical Reviews* 106: 2299–319.

Adams, M.R. 2014. "Vinegar." In *Encyclopedia of Food Microbiology*, 2nd ed., edited by Carl A. Batt and Richard K. Robinson. London: Academic Press.

Adjei, Maame Y.B. 2017. "Applications and Limitations of Discrimination Testing." In *Discrimination Testing in Sensory Science: A Practical Handbook*, edited by Lauren Rogers, 85–105. Duxford, UK: Woodhead Publishing.

ASBC. *ASBC Methods of Analysis*, online. 14th ed. St. Paul, MN: American Society of Brewing Chemists. https://www.asbcnet.org/Methods/ (subscription required).

Bachmanov, Alexander A. and Gary K. Beauchamp. 2007. "Taste Receptor Genes." *Annual Review of Nutrition* 27: 389–414.

Banerjee Amitav, U. B. Chitnis, S.L. Jadhav, J.S. Bhawalkar, and S. Chaudhury. 2009. "Hypothesis testing, type I and type II errors." *Industrial Psychiatry Journal* 18(2): 127–31.

Boulton, Chris and David Quain. 2001. *Brewing Yeast and Fermentation*. Oxford: Blackwell Science Ltd.

Brown, Perter C., Henry L. Roediger (III), and Mark A. McDaniel. 2014. *Make It Stick: The Science of Successful Learning*. Cambridge, MA: Harvard University Press.

Buehler, Roger, Dale Griffin, and Michael Ross. 1994. "Exploring the 'Planning Fallacy': Why People Underestimate Their Task Completion Times." *Journal of Personality and Social Psychology* 67(3): 366–81.

Buss, David M. 2005. *The Handbook of Evolutionary Psychology*. 1st ed. Hoboken, NJ: John Wiley and Sons.

Cantwell, Dick and Peter Bouckaert. 2016. *Wood and Beer: A Brewer's Guide*. Boulder, CO: Brewers Publications.

Cilurzo, Vincent. 2012. "Sour Beer." In *The Oxford Companion to Beer*, edited by Garrett Oliver. Oxford: Oxford University Press. Kindle.

Collings, Virginia B. 1974. "Human taste response as a function of locus of stimulation on the tongue and soft palate." *Perception & Psychophysics* 16: 169–74.

Croy, I., S. Olgun, L. Mueller, A. Schmidt, M. Muench, G. Gisselmann, H. Hatt, and T. Hummel. 2016. "Spezifische Anosmie als Prinzip olfaktorischer Wahrnehmung [Specific Anosmia as a Principle of Olfactory Perception]." *HNO* 64(5): 292–95.

Danziger, Shai, Jonathan Levav, and Liora Avnaim-Pesso. 2011. "Extraneous factors in judicial decisions." *Proceedings of the National Academy of Sciences of the United States of America.* 108(17): 6889–92.

De Keukeleire, Denis, Arne Heyerick, Kevin Huvaere, Leif H. Skibsted, and Mogens L. Andersen. 2008. "Beer lightstruck flavor: The full story." *Cerevisia* 33(3): 133–44.

Dean, M. L. 1980. "Presentation order effects in product taste tests." *Journal of Psychology* 105: 107–10.

Dodge, Yadolah. 2008. "Gosset, William Sealy." In *The Concise Encyclopedia of Statistics,* 234–235. New York: Springer-Verlag.

Doty, Richard L. 2015. *Handbook of Olfaction and Gustation.* 3rd ed. Hoboken, NJ: John Wiley & Sons.

Fix, George. 1999. *Principles of Brewing Science: A Study of Serious Brewing Issues.* 2nd ed. Boulder, CO: Brewers Publications.

Ghasemi-Varnamkhasti, Mahdi, Seyed Saeid Mohtawebi, Maryam Siadat, Jesus Lozano, Hojat Ahmadi, Seyed Hadi Razavi, and Amadou Dicko. 2011. "Aging fingerprint characterization of beer using electronic nose." *Sensor and Actuators B: Chemical* 159: 51–59.

Glindemann, Dietmar, Andrea Dietrich, Hans-Joachim Staerk, and Peter Kuschk. 2006. "The Two Odors of Iron when Touched or Pickled: (Skin) Carbonyl Compounds and Organophosphines." *Angewandte Chemie* 45: 7006–9.

Grbin, Paul R., Markus Herderich, Andrew Markides, Terry H. Lee, and Paul A. Henschke. 2007. "The Role of Lysine Amino Nitrogen in the Biosynthesis of Mousy Off-Flavor Compounds by *Dekkera anomala*." *Journal of Agricultural and Food Chemistry* 55: 10872–79.

Griffin, Ricky W. 2013. *Fundamentals of Management.* 8th ed. Boston: Cengage Learning.

Hampson, Tim. 2012. "Burton Snatch." In *The Oxford Companion to Beer*, edited by Garrett Oliver. Oxford: Oxford University Press. Kindle.

Harwood, Meriel L., Joseph R. Loquasto, Robert F. Roberts, Gregory R. Ziegler, and John E. Hayes. 2013. "Explaining tolerance for bitterness in chocolate ice cream using solid chocolate preferences." *Journal of Dairy Science* 96(8): 4938–44.

Hewson L., T. Hollowood, S. Chandra, and J. Hort. 2009. "Gustatory, Olfactory and Trigeminal Interactions in a Model Carbonated Beverage." *Chemosensory Perception* 2: 94–107.

Inoue, Takashi. 2008. *Diacetyl in Fermented Foods and Beverages.* St. Paul, MN: American Society of Brewing Chemists.

Jiang, Peihua, Jesusa Josue, Xia Li, Dieter Glaser, Weihua Li, Joseph G. Brand, Robert F. Margolskee, Danielle R. Reed, and Gary K. Beauchamp. 2012. "Major taste loss in carnivorous mammals." *Proceedings of the National Academy of Sciences of the United States of America* 109(13): 4956–61.

Joesten, Melvin, Mary E. Castellion, and John L. Hogg. 2007. *The World of Chemistry: Essentials*. 4th Ed. Belmont, CA: Thomson Brooks/Cole.

Johansen-Berg, Heidi, and Donna M Lloyd. 2000. "The Physiology and Psychology of Selective Attention to Touch." *Frontiers in Bioscience* 5: 894–904.

Kahneman, Daniel. 2011. *Thinking, Fast and Slow*. New York: Farrar, Straus, and Giroux. Kindle.

Keast, Russell SJ and Andrew Costanzo. 2015. "Is fat the sixth taste primary? Evidence and implications." *Flavour* 4: 5.

Keast, Russell SJ, Paul A.S. Breslin, and Gary K. Beauchamp. 2001. "Suppression of Bitterness Using Sodium Salts." *CHIMIA* 55(5): 441–47.

Kilcast, David, ed. 2010. *Sensory Analysis for Food and Beverage Quality Control: A Practical Guide*. Cambridge: Woodhead Publishing.

Kolb, Rachel R. and Marcey L. Hoover. 2012. *The History of Quality in Industry*. SAND2012-7060. August 2012. Albuquerque, NM: Sandia National Laboratories.

Kraus-Weyermann, Thomas. 2012. "Malt." In *The Oxford Companion to Beer*, edited by Garrett Oliver. Oxford: Oxford University Press. Kindle.

Kubiak, T. M. and Donald W. Benbow. 2016. *The Certified Six Sigma Black Belt Handbook*. 3rd ed. Milwaukee, WI: ASQ Quality Press.

Langstaff, Susan A., J.-X. Guinard, and M. J. Lewis. 1991. "Instrumental Evaluation of the Mouthfeel of Beer and Correlation with Sensory Evaluation." *Journal of the Institute of Brewing* 97: 427–33.

Langstaff, Susan A. and M. J. Lewis. 1993. "The Mouthfeel of Beer – A Review." *Journal of the Institute of Brewing* 99: 31–37.

Lawless, Harry T. and Hildegarde Heymann. 2010. *Sensory Evaluation of Food: Principles and Practices*. 2nd ed. New York: Springer Science+Business Media, LLC.

Lermusieau, G., S. Noël, C. Liégeois, and S. Collin. 1999. "Nonoxidative Mechanism for Development of *trans*-2-Nonenal in Beer." *Journal of the American Society of Brewing Chemists* 57(1): 29–33.

Lipscomb, Keri, James Rieck, and Paul Dawson. 2016. "Effect of Temperature on the Intensity of Basic Tastes: Sweet, Salty and Sour." *Journal of Food Research* 5(4): 1–10.

Mantonakis, Antonia, Pauline Rodero, Isabelle Lesschaeve, and Reid Hastie. 2009. "Order in Choice: Effects of Serial Position on Preferences." *Psychological Science* 20(11): 1309–12.

McQuaid, John. 2015. *Tasty: The Art and Science of What We Eat*. New York: Scribner.

Meilgaard, M. C., C. E. Dalgliesh, and J. F. Clapperton. 1979. "Beer Flavor Terminology." *Journal of the American Society of Brewing Chemists*, 37(1): 47–52.

Meilgaard, M. C., Gail Vance Civille, and B. Thomas Carr. 2016. *Sensory Evaluation Techniques.* 5th ed. Boca Raton, FL: CRC Press, Taylor & Francis Group.

Miller, Jeff. 2016. "Hypothesis Testing in the Real World." *Educational and Psychological Measurement* 77(4): 663–72.

Miyazaki, Masao, Tetsuro Yamashita, Yusuke Suzuki, Yoshihiro Saito, Satoshi Soeta, Hideharu Taira, and Akemi Suzuki. 2006. "A Major Urinary Protein of the Domestic Cat Regulates the Production of Felinine, a Putative Pheromone Precursor." *Chemistry and Biology* 13: 1071–79.

Montgomery, Douglas C. 2009. *Introduction to Statistical Quality Control.* 6th ed. Hoboken, NJ: John Wiley & Sons, Inc.

Moore, John E. A., L. Janet Forrester, and Paolo Pelosi. 1976. "Specific Anosmia to Isobutyraldehyde: The Malty Primary Odor." *Chemical Senses* 2(1): 17–25.

Morrot, Gil, Frédéric Brochet, and Denis Dubourdieu. 2001. "The Color of Odors." *Brain and Language* 79: 309–320.

Muñoz, Alejandra M., Gail Vance Civille, and B. Thomas Carr. 1992. *Sensory Evaluation in Quality Control.* New York: Van Nostrand Reinhold.

Nisbett, Richard E. and Timothy DeCamp Wilson. 1977. "The Halo Effect: Evidence for Unconscious Alteration of Judgments." *Journal of Personality and Social Psychology* 35(4): 250–56.

Norton, Michael I., Daniel Mochon, and Dan Ariely. 2011. "The IKEA effect: When labor leads to love." *Journal of Consumer Psychology* 22(3): 453–60.

Obayashi, Yoko and Yoichi Nagamura. 2016. "Does monosodium glutamate really cause headache? A systematic review of human studies." *Journal of Headache and Pain* 17: 54.

Oladokun, Olayide, Amparo Tarrega, Sue James, Trevor Cowley, Frieda Dehrmann, Katherine Smart, David Cook, and Joanne Hort. 2016. "Modification of perceived beer bitterness intensity, character and temporal profile by hop aroma extract." *Food Research International* 86: 104–11.

O'Mahony, M., M. Goldenberg, J. Stedmon, and J. Alford. 1979. "Confusion in the use of the taste adjectives 'sour' and 'bitter'." *Chemical Senses* 4(4): 301–18.

Pellettieri, Mary. 2015. *Quality Management: Essential Planning for Breweries.* Boulder, CO: Brewers Publications.

Philliskirk, George. 2012. "Esters." In *The Oxford Companion to Beer*, edited by Garrett Oliver. Oxford: Oxford University Press. Kindle.

Plotto, A., K. W. Barnes, and K. L. Goodner. 2006. "Specific Anosmia Observed for β-Ionone, but not for α-Ionone: Significance for Flavor Research." *Journal of Food Science* 71(5): 401–6.

Rettberg, Nils, Martin Biendl, and Leif-Alexander Garbe. 2018. "Hop Aroma and Hoppy Beer Flavor: Chemical Backgrounds and Analytical Tools—A Review." *Journal of the American Society of Brewing Chemists* 76(1): 1–20.

Rogers, L.L. 2010. "Using sensory techniques for shelf-life assessment." In *Sensory Analysis for Food and Beverage Quality Control: A Practical Guide*, edited by David Kilcast, 143–55. Duxford, UK: Woodhead Publishing.

Rogers, Lauren, ed. 2017. *Discrimination Testing in Sensory Science: A Practical Handbook*. Duxford: Woodhead Publishing.

Saison, Daan, David P. De Schutter, Bregt Uyttenhove, Filip Delvaux, and Freddy R. Delvaux. 2009. "Contribution of staling compounds to the aged flavour of lager beer by studying their flavour thresholds." *Food Chemistry* 114(4): 1206–15.

Sanderson, Tracey. 2017. "Tetrad Test." In *Discrimination Testing in Sensory Science: A Practical Handbook*, edited by Lauren Rogers, 183–95. Duxford, UK: Woodhead Publishing.

Sarafoleanu, C., C. Mella, M. Georgescu, and C. Perederco. 2009. "The importance of the olfactory sense in the human behavior and evolution." *Journal of Medicine and Life* 2(2): 196–98.

Scopelliti, Irene, Carey K. Morewedge, Erin McCormick, H. Lauren Min, Sophie Lebrecht, and Kim S. Kassam. 2015. "Bias Blind Spot: Structure, Measurement, and Consequences." *Management Science* 61(10): 2468–86.

Shepherd, Gordon M. 2012. *Neurogastronomy: How the Brain Creates Flavor and Why It Matters.* New York: Columbia University Press.

Smythe, John E. and Charles W. Bamforth. 2000. "Shortcomings in Standard Instrumental Methods for Assessing Beer Color." *Journal of the American Society of Brewing Chemists* 58(4): 165–66.

Snowdon, Eleanor M., Michael C. Bowyer, Paul R. Grbin, and Paul K. Bowyer. 2006. "Mousy Off-Flavor: A Review." *Journal of Agricultural and Food Chemistry* 54: 6465–74.

Sparrow, Jeff. 2005. *Wild Brews: Beer Beyond the Influence of Brewer's Yeast.* Boulder, CO: Brewers Publications.

Spence, Charles, Carmel A. Levitan, Maya U. Shankar, and Massimiliano Zampini. 2010. "Does Food Color Influence Taste and Flavor Perception in Humans?" *Chemosensory Perception* 3: 68–84.

Spence, Charles. 2012. "Auditory contributions to flavour perception and feeding behaviour." *Physiology and Behavior* 107(4): 505–15.

Steinhaus, Martin and Peter Schieberle. 2000. "Comparison of the Most Odor-Active Compounds in Fresh and Dried Hop Cones (*Humulus lupulus* L. variety Spalter Select) Based on GC-Olfactometry and Odor Dilution Techniques." *Journal of Agricultural and Food Chemistry* 48: 1776–83.

Stewart, G.G., and I. Russell. 1998. *An Introduction to Brewing Science & Technology: Series III; Brewer's Yeast.* London: Institute of Brewing.

Stone, Herbert and Joel L. Sidel. 2004. *Sensory Evaluation Practices.* 3rd ed. San Diego: Elsevier Academic Press.

Storgårds, Erna. 2000. "Process Hygiene Control in Beer Production and Dispensing." Academic dissertation, University of Helsinki. VTT Publications 410. Published by Technical Research Center of Finland (VTT), Espoo, Finland. 105 p.

Talavera, K., Y. Ninomiya, C. Winkel, T. Voets, and B. Nilius. 2007. "Influence of temperature on taste perception." *Cellular and Molecular Life Sciences* 64: 377–81.

Tchobanov, Iavor, Laurent Gal, Michèle Guilloux-Benatier, Fabienne Remize, Tiziana Nardi, Jean Guzzo, Virginie Serpaggi, and Hervé Alexandre. 2008. "Partial vinylphenol reductase purification and characterization from *Brettanomyces bruxellensis.*" *FEMS Microbiology Letters* 284(2): 213–17.

Theerasilp, Sarroch and Yoshie Kurihara. 1988. "Complete Purification and Characterization of the Taste-modifying Protein, Miraculin, from Miracle Fruit." *Journal of Biological Chemistry* 263(23): 11536–39.

Tonsmeire, Michael. 2014. *American Sour Beers: Innovative Techniques for Mixed Fermentations.* Boulder, CO: Brewers Publications.

Vanderhaegen, Bart, Hedwig Neven, Hubert Verachtert, and Guy Derdelinckx. 2006. "The chemistry of beer aging — a critical review." *Food Chemistry* 95: 357–81.

Vaughn, Michael. 2013. *The Thinking Effect: Rethinking Thinking to Create Great Leaders and the New Value Worker.* Boston: Nicholas Brealey Publishing.

Vriesekoop, Frank, Moritz Krahl, Barry Hucker, and Garry Menz. 2012. "125th Anniversary Review: Bacteria in brewing; The good, the bad and the ugly." *Journal of the Institute of Brewing and Distilling* 118: 335–45.

Wheeler, Donald J. 2010. *Understanding Statistical Process Control.* 3rd ed. Knoxville, TN: SPC Press.

White, Chris and Jamil Zainasheff. 2010. *Yeast: The Practical Guide to Beer Fermentation.* Boulder, CO: Brewers Publications.

Wiener, Ayana, Mariana Shudler, Anat Levit, and Masha Y. Niv. 2012. "BitterDB: a database of bitter compounds." *Nucleic Acids Research* 40: 413–19.

Wong, Bang. 2011. "Points of view: Color blindness." *Nature Methods* 8: 441.

Yakobson, Chad Michael. 2012. "Brettanomyces." In *The Oxford Companion to Beer*, edited by Garrett Oliver. Oxford: Oxford University Press. Kindle.

Ye, Wenlei, Rui B. Chang, Jeremy D. Bushman, Yu-Hsiang Tu, Eric M. Mulhall, Courtney E. Wilson, Alexander J. Cooper, Wallace S. Chick, et al. 2015. "The K$^+$ channel K$_{IR}$2.1 functions in tandem with proton influx to mediate sour taste transduction." *Proceedings of the National Academy of Sciences of the United States of America* 113(2): E229–E238.

Zampini, Massimiliano and Charles Spence. 2005. "The Role of Auditory Cues in Modulating the Perceived Crispness and Staleness of Potato Chips." *Journal of Sensory Studies* 19(5): 347–63.

Zhang, Xiaohong and Stuart Firestein. 2007. "Nose thyself: individuality in the human olfactory genome." *Genome Biology* 8(11): 230.

INDEX